Extraction Metallurgy - New Perspectives

Edited by Swamini Chopra and Thoguluva Vijayaram

Published in London, United Kingdom

Extraction Metallurgy - New Perspectives
http://dx.doi.org/10.5772/intechopen.102110
Edited by Swamini Chopra and Thoguluva Vijayaram

Contributors
Prasunjit Sengupta, Nnaemeka Stanislaus Nzeh, Maite Mokgalaka, Nthabiseng Maila, Patricia Popoola, Daniel Okanigbe, Abraham Adeleke, Samson Adeosun, Peter Futas, Alena Pribulova, Sneha Dandekar, Dilip Peshwe, Kavita Pande, Swamini Chopra, Olga Mokhodoeva

First published in London, United Kingdom, 2024 by IntechOpen
IntechOpen is the global imprint of INTECHOPEN LIMITED, registered in England and Wales, registration number: 11086078, 5 Princes Gate Court, London, SW7 2QJ, United Kingdom

British Library Cataloguing-in-Publication Data
A catalogue record for this book is available from the British Library

Additional hard and PDF copies can be obtained from orders@intechopen.com

Extraction Metallurgy - New Perspectives
Edited by Swamini Chopra and Thoguluva Vijayaram
p. cm.
Print ISBN 978-1-83969-262-8
Online ISBN 978-1-83969-263-5
eBook (PDF) ISBN 978-1-83969-264-2

For EU product safety concerns:
IN TECH d.o.o., Prolaz Marije Krucifikse Kozulić 3, 51000 Rijeka, Croatia,
info@intechopen.com or visit our website at intechopen.com.

Meet the editors

Dr. Swamini Chopra is an assistant professor at the Maharashtra Institute of Technology (MIT), India, working in materials engineering at the Center of Excellence in Materials and Metallurgy. She was formerly associated with the Defence Institute of Advanced Technology (DIAT), India, on a project sponsored by the Department of Science and Technology (DST), Government of India. Dr. Chopra has been recognized twice by the Cryogenic Society of America for her contributions to the field of cryogenics. She has more than thirty journal publications and three patents to her credit. She has also authored two books and edited two books on the topics of materials science, carbon-based nanomaterials, and cutting tools. She is actively associated with undergraduate, postgraduate, and Ph.D. scholars at her institute and her areas of interest include cryogenic treatment, polymer composites, applications of carbon-based materials, and mechanical performance of materials.

Dr. T. R. Vijayaram is a Senior Professor in the Department of Mechanical Engineering under the School of Mechanical Engineering, BIST, Bharath Institute of Higher Education and Research (BIHER), Selaiyur, Chennai, Tamil Nadu, India. He is an expert in the field of materials and metallurgical, manufacturing, and mechanical engineering. He received his B.E. degree in Mechanical Engineering from Madurai Kamaraj University, followed by an M.E. degree in Industrial Metallurgy from PSG College of Technology, Bharathiyar University. He obtained his Ph.D. research degree in Mechanical Engineering from Universiti Putra Malaysia, Malaysia. Later, he worked as a Rector researcher in metallurgy at DCCI, Genoa University, Italy. His passion for academics, research, and education led him to obtain an MBA in Educational Management and an MA in Sociology from the University of Madras. He is also a Chartered Engineer (India) and a member of several professional and scientific bodies in India and abroad like ISTE (Life Member), IEI, IIF, and SAE (USA). He is also a Fellow of the Institution of Engineers, FIE, India. Recently, Dr. T. R. Vijayaram received the Distinguished Scientist Award in Metallurgical and Materials Engineering for his outstanding contribution to metallurgy. He has published more than 230 papers in international and national journals, conferences, broadsheets, and magazines. His areas of research include materials engineering, metallurgical engineering, manufacturing engineering, and mechanical engineering.

Contents

Preface

In the ever-evolving realm of metallurgy, the pursuit of innovative methods and new perspectives is the crucible in which progress is forged. *Extraction Metallurgy – New Perspectives* represents a milestone in our collective journey to understand and harness the intricate processes that transform ores and minerals from the Earth into the materials that underpin our modern world. As the editor of this volume, I am both honored and excited to present a comprehensive exploration of the latest advancements and fresh insights in this dynamic field. Metallurgy, often described as the science and art of extracting metals from their ores and shaping them into useful forms, has a rich history dating back millennia. From the dawn of civilization to the present day, metallurgy has been instrumental in shaping human society, powering industrial revolutions, and fueling technological innovations. Yet, as we stand on the precipice of a new era marked by sustainability and resource consciousness, the traditional paradigms of metallurgy are being redefined. This book delves into the heart of modern extraction metallurgy, where traditional knowledge intersects with cutting-edge research. It covers a wide array of topics, from the fundamentals of ore characterization to the complexities of hydrometallurgical processes. It explores the environmental and economic aspects of mineral extraction and presents novel approaches to reduce the environmental footprint of the industry.

Extraction Metallurgy – New Perspectives embraces the interdisciplinary nature of modern metallurgy. Advances in materials science, chemistry, physics, and engineering converge in the pursuit of more efficient and environmentally responsible extraction processes. Researchers and practitioners from diverse fields have contributed their expertise to this volume, emphasizing the importance of collaboration in pushing the boundaries of metallurgical knowledge. Moreover, this book is not confined to the laboratory or industrial setting. It recognizes the global significance of mineral resources and the critical role they play in geopolitics and economic development. By examining the geopolitical dynamics of mineral extraction and the ethical considerations surrounding resource distribution, this book invites readers to reflect on the broader implications of our metallurgical endeavors.

In closing, I would like to express my profound gratitude to the contributors who have shared their insights, experiences, and expertise to make this book a reality. I extend my thanks to my colleagues, mentors, and the entire metallurgical community for their unwavering dedication to advancing our field. And I invite readers, whether seasoned professionals or curious novices, to embark on this journey through *Extraction Metallurgy – New Perspectives*, where tradition meets innovation and where the future of our industry is being shaped. May this volume serve as a source of inspiration, knowledge, and collaboration as we forge ahead in the pursuit of sustainable, responsible, and transformative metallurgy.

Swamini Chopra
Center of Excellence in Materials and Metallurgy,
Maharashtra Institute of Technology,
Aurangabad (M.S.), India

Thoguluva Vijayaram
Senior Professor,
Department of Mechanical Engineering,
School of Mechanical Engineering,
BIST, BIHER, Bharath Institute of Higher Education and Research,
Selaiyur, Chennai, Tamil Nadu, India

Section 1

Introduction

Chapter 1

Introductory Chapter: Extraction Metallurgy – New Perspectives

Swamini Chopra

1. Introduction

Metallurgy is the scientific field that deals with the study of metals and their properties. While extraction metallurgy is a branch of metallurgical engineering that focuses on the extraction of metals from ores, their refining techniques and converting them into useful materials. Extraction of useful metals is a vital aspect of metallurgy, and nowadays it is not only limited to metallic ores [1, 2]. The mining and refining of minerals like borosilicate and calcium carbonate for engineering applications is also considered a part of extraction science [3]. The extraction process involves a series of complex steps that require specialized knowledge, advanced technology, and highly skilled workers. Extraction metallurgy is critical to modern society, as it provides the raw materials for many industries, including aerospace, construction, energy, and electronics. This book brings together some of the recent developments in the field of extraction and refining of extracted products, followed by their applications. Meanwhile, this introductory chapter will provide an overview about extraction metallurgy, including its history, importance, and the main steps involved in the process. The chapter will also highlight some of the key challenges and research advances in extraction metallurgy, as well as some of the environmental and social impacts of the industry.

1.1 History of extraction metallurgy

The history of extraction metallurgy can be traced back thousands of years, to the early days of human civilization. The earliest metalworkers used simple techniques to extract metals from their ores, such as heating them in fires or washing them with water. Over time, as metallurgical knowledge and technology advanced, more complex methods were developed [4, 5].

One of the most important advances in extraction metallurgy occurred during the industrial revolution in the eighteenth and nineteenth centuries [6]. During this time, new technologies were developed that enabled large-scale production of metals such as iron and steel. The Bessemer process, invented in the 1850s, was one of the most significant advances in the field. This process involved blowing air through molten iron to remove impurities and produce high-quality steel [7, 8].

Since then, extraction metallurgy has continued to evolve and improve, with new technologies and methods being developed to extract a wider range of products and produce them more efficiently.

2. Importance of extraction metallurgy

The use of metals and minerals is an essential component of modern society. Metals are used in the construction of buildings, transportation systems, and consumer goods such as electronics and appliances [9]. Minerals are used in a wide range of industries, including agriculture, construction, and manufacturing [10, 11]. However, the raw materials for these products are not readily available in their usable forms. The ores that contain these metals and minerals are often found in remote locations and require extensive processing to extract the valuable metals and minerals.

Extraction metallurgy is the process of extracting metals and minerals from their ores. It is a critical process in the mining industry and plays an essential role in the production of many industrial and consumer goods [2]. The extraction of metals involves a series of steps that are designed to separate the desired metal from the rest of the ore. This process involves the use of various chemical, physical, and mechanical methods to separate the metal from the ore. Extraction metallurgy is an important process that has significant economic and environmental implications. The economic importance of extraction metallurgy is evident from the fact that it is used to extract valuable metals that are used in various industrial processes [12]. These metals include copper, gold, silver, lead, zinc, nickel, and iron. The value of these metals varies, but they are all in demand due to their unique properties and uses. For example, copper is one of the most widely used metals in the world and is used in electrical wiring, plumbing, and electronic devices. Iron is used in the construction of buildings, bridges, and vehicles. Zinc is used in galvanizing steel, and aluminum is used in the construction of airplanes, cars, and beverage cans. Extraction metallurgy has also played a critical role in the development of new technologies. For example, the extraction of lithium from brines has become essential for the production of lithium-ion batteries [13, 14], which are used in electric vehicles and portable electronic devices. The extraction of rare earth elements from ores has also become increasingly important for the production of electronics and advanced technologies.

Extraction metallurgy also plays a crucial role in reducing the environmental impact of mining. Mining can have significant environmental consequences, including soil erosion, water pollution, and habitat destruction. By using extraction metallurgy to extract metals from ores, the environmental impact of mining can be reduced. This is because extraction metallurgy involves the use of various chemical and physical processes that allow for the efficient extraction of products from ores.

3. Importance of research in extraction metallurgy

Research in extraction metallurgy is important because it allows for the development of new and more efficient processes for extracting metals and minerals from ores. This is important because their demand is ever increasing, and traditional extraction processes may not be able to keep up with this demand. Research in extraction metallurgy also allows for the development of more environmentally friendly processes that reduce the impact of mining on the environment. For example, investigating the use of bioleaching to extract copper from its ore is proving to be an efficient process by using microorganisms to break down the sulfide minerals in copper ore [15, 16]. This process might prove to be more environmentally friendly than traditional smelting as it produces fewer emissions and requires less energy.

Research in extraction metallurgy also allows for the development of new materials and alloys that have unique properties and uses. In near future, the use of nanotechnology to develop new materials with unique properties will accelerate the future of metallurgy. These materials may have applications in various industries, including electronics, medicine, and energy.

4. Challenges in extraction metallurgy

Apart from the benefits, extraction metallurgy is a complex process that involves many challenges. One of the biggest challenges is the fact that ores are often complex mixtures of minerals that must be separated to extract the desired component [17, 18]. This requires the use of various chemical and physical processes that can be expensive and time-consuming. Another challenge in extraction metallurgy is the fact that ores may contain impurities that can interfere with the extraction process. For example, copper ore may contain sulfur, which can form sulfur dioxide during smelting [15]. This can lead to environmental problems such as acid rain.

5. Extraction metallurgy process

Irrespective of the metal or mineral to be extracted, the process of extraction remains fairly same. It involves steps like mining, crushing, grinding, concentration, smelting, refining, and casting. The choice of method for extraction depends on several factors such as the type of ore, the concentration of the metal or mineral within the ore, and the desired purity of the final product. These steps are discussed below:

- Mining

The first step in extraction is the mining. Mining involves the extraction of ores from the earth's crust. Ores are rocks that contain valuable metals such as copper, gold, silver, and platinum. The mining process involves drilling, blasting, and hauling the ore to the surface.

- Crushing and grinding

Once the ore is extracted, it is crushed and ground into small particles. This process increases the surface area of the ore, making it easier to extract the metal or mineral. The crushing and grinding process may involve several stages, depending on the size of the ore and the desired particle size.

- Concentration

After the ore is crushed and ground, it is concentrated to increase the concentration of the metal or mineral to be extracted. Concentration involves separating the desired material from the unwanted components present in the ore. This is typically done using gravity separation, flotation, or magnetic separation. Gravity separation involves using the differences in the density of the metal or mineral and the surrounding rock to separate them. Flotation involves using chemicals to make the metal or mineral hydrophobic, or repel water, and the surrounding rock hydrophilic, or attract water. Magnetic separation involves using magnets to separate the metal or mineral from the surrounding rock.

- Smelting

Smelting is the process of extracting metal from its ore by heating it to a high temperature in the presence of a reducing agent. The reducing agent reduces the metal oxide to metal. The most commonly used reducing agent is coke (a form of carbon). The metal is then separated from the slag (the waste material) using various techniques.

- Refining

After smelting, the metal is purified further to remove any impurities. This process is known as refining. This process is used for minerals as well and involves the use of various techniques such as electrolysis, precipitation, and distillation.

- Casting

The final step in metal extraction is casting. The purified metal is cast into various shapes and forms, depending on its intended use. Casting involves melting the metal and pouring it into a mold to give it the desired shape. This step is not usually needed for the minerals.

6. Extraction techniques

The extraction of useful products can be achieved by following techniques [19, 20]:

- Hydrometallurgy

Hydrometallurgy is a metal extraction technique that involves the use of aqueous solutions to extract metals from their ores. In this technique, the metal is dissolved in an acidic or alkaline solution, and the impurities are separated using various methods.

- Pyrometallurgy

Pyrometallurgy is a metal extraction technique that involves the use of high temperatures to extract metals from their ores. In this technique, the ore is heated to high temperatures, and the metal is extracted using various methods.

- Electrometallurgy

Electrometallurgy is a metal extraction technique that involves the use of electricity to extract metals from their ores. In this technique, the metal ions in the ore are reduced to metal using an electric current.

DOI: http://dx.doi.org/10.5772/intechopen.112882

Author details

Swamini Chopra
Centre of Excellence in Materials and Metallurgy, Maharashtra Institute of Technology, Chhatrapati Sambhajinagar, Maharashtra, India

*Address all correspondence to: chopra.swamini@gmail.com

References

[1] Lu L, Pan J, Zhu D. 16—Quality requirements of iron ore for iron production [Internet]. In: Lu L, editor. Iron Ore. Cambridge: Woodhead Publishing, Elsevier; 2015. pp. 475-504. Available from: https://www.sciencedirect.com/science/article/pii/B9781782421566000162

[2] Whitworth AJ, Vaughan J, Southam G, van der Ent A, Nkrumah PN, Ma X, et al. Review on metal extraction technologies suitable for critical metal recovery from mining and processing wastes. Minerals Engineering. 2022;**182**:107537

[3] Francis LF. Materials Processing: A Unified Approach to Processing of Metals, Ceramics and Polymers. Cambridge: Academic Press, Elsevier; 2015

[4] Forrester R. History of Metallurgy [Internet]. 2016. Available from: https://papers.ssrn.com/abstract=2864178

[5] Kozina F, Zovko BZ. Metallurgy development: Discovery and utilization of aluminum through history. Engineering Power: Bulletin of the Croatian Academy of Engineering. 2022;**17**:6-19

[6] Mokyr J, Strotz RH. The Second Industrial Revolution, 1870-1914. Evanston IL: Northwestern University; 2003. Available from: https://faculty.wcas.northwestern.edu/jmokyr/castronovo.pdf

[7] Ghosh A, Chatterjee A. Iron Making and Steelmaking: Theory and Practice. India: PHI Learning Pvt. Ltd.; 2008

[8] Lv M, Chen S, Yang L, Wei G. Research progress on injection technology in converter steelmaking process. Metals. 2022;**12**:1918

[9] Hertwich EG, Ali S, Ciacci L, Fishman T, Heeren N, Masanet E, et al. Material efficiency strategies to reducing greenhouse gas emissions associated with buildings, vehicles, and electronics—a review. Environmental Research Letters. 2019;**14**:043004

[10] Murray HH. Traditional and new applications for kaolin, smectite, and palygorskite: A general overview. Applied Clay Science. 2000;**17**:207-221

[11] Franks DM. Reclaiming the neglected minerals of development. The Extractive Industries and Society. 2020;**7**:453-460

[12] Ayres RU. Metals recycling: Economic and environmental implications. Resources, Conservation and Recycling. 1997;**21**:145-173

[13] Meng F, McNeice J, Zadeh SS, Ghahreman A. Review of lithium production and recovery from minerals, brines, and lithium-ion batteries. Mineral Processing and Extractive Metallurgy Review. 2021;**42**(2):123-141. Available from: https://www.tandfonline.com/doi/abs/10.1080/08827508.2019.1668387

[14] Bae H, Kim Y. Technologies of lithium recycling from waste lithium ion batteries: A review materials advances. DOI: 10.1039/D1MA00216C. Available from: https://pubs.rsc.org/en/content/articlehtml/2021/ma/d1ma00216c

[15] Watling HR. The bioleaching of sulphide minerals with emphasis on copper sulphides—A review. Hydrometallurgy. 2006;**84**:81-108

[16] Sajjad W, Zheng G, Din G, Ma X, Rafiq M, Xu W. Metals extraction from sulfide ores with microorganisms: The bioleaching technology and recent

developments. Transactions of the Indian Institute of Metals. 2019;**72**:559-579

[17] Izatt RM, Izatt SR, Bruening RL, Izatt NE, Moyer BA. Challenges to achievement of metal sustainability in our high-tech society. Chemical Society Reviews. 2014;**43**:2451-2475

[18] Sarker SK, Haque N, Bhuiyan M, Bruckard W, Pramanik BK. Recovery of strategically important critical minerals from mine tailings. Journal of Environmental Chemical Engineering. 2022;**10**:107622

[19] Trinh HB, Chun LJ, Jae SY, Lee J. A review on the recycling processes of spent auto-catalysts: Towards the development of sustainable metallurgy. Waste Management. 2020;**114**:148-165

[20] Mizuno N, Kosai S, Yamasue E. Microwave-based extractive metallurgy to obtain pure metals: A review. Cleaner Engineering and Technology. 2021;**5**:100306

Section 2

Extraction Science

Chapter 2

Pyrometallurgical Approach in the Recovery of Niobium and Tantalum

Nnaemeka Stanislaus Nzeh, Maite Mokgalaka, Nthabiseng Maila, Patricia Popoola, Daniel Okanigbe, Abraham Adeleke and Samson Adeosun

Abstract

The pyrometallurgical methods of the recovery of various critical metals have been established. Certain pyrometallurgical approaches for niobium (Nb) and tantalum (Ta) extraction have been studied and investigated by several researchers. For instance, the conventional reduction of Nb mineral or Nb_2O_5 to Nb metal has been conducted over the decades. Albeit, the success level of this process, it however involves the consumption of lots of energy, high cost of equipment/equipment maintenance, somewhat low Nb and Ta recovery and limited effectiveness on low grade minerals; and thus, considered cost intensive and inefficient. In addition, the inadequacies of pyrometallurgical extraction of these metals from their complex, low grade mineral ores due to its difficulty and large energy requirement in melting the elemental impurities and gangue minerals have been a major concern. On this premise therefore, the study will provide insights into recent pyrometallurgical techniques of Nb and Ta extraction as well as associated factors and challenges.

Keywords: niobium, tantalum, mineral ore, reduction, roasting, fusion, precipitation, high temperature, pyrometallurgical, extraction, decomposition, beneficiation, separation, recovery

1. Introduction

The extraction and purification of metals from their mineral complex or concentrates, based on physicochemical changes occurring at elevated/somewhat high temperatures, is referred to as pyrometallurgy. Pyrometallurgical processes essentially involves certain heating procedures; usually employing dry methods conducted at high or elevated temperatures and may also involve the melting of the charge or feed material as well as high-temperature processes in which chemical reactions occur between gases, solids, and molten materials. Mineral solids composed of valuable metals are processed to make intermediate compounds for further processing or to their elemental or metallic state. Calcining, roasting and smelting operations are typical pyro-metallurgical processes that involve gases and solids, and may also result in molten products. The exothermic character of the chemical reactions occurring may be the source of the energy needed to sustain the high temperature pyro-metallurgical processes. Most frequently, fuel is being used to contribute energy to the process, or in the case of some smelting procedures, electrical energy is being applied directly. Roasting on the other hand may however involve thermal gas–solid processes, such as oxidation, reduction, chlorination, sulphation, and pyro-hydrolysis [1]. The pyrometallurgical methods of the recovery of various critical metals have been established over the decades. Certain pyrometallurgical approaches for niobium (Nb) and tantalum (Ta) extraction have been studied and investigated by several researchers in various parts of the world. For instance, the conventional reduction practice of Nb/Ta minerals or Nb_2O_5/Ta_2O_5 to Nb/Ta metals has been conducted for several decades. Albeit, the measure of success of this process, pyrometallurgical procedure was reported to involve high temperatures and high energy consumption, high equipment cost, use of sophisticated equipment and high cost of (equipment) maintenance, somewhat low Nb and Ta yield/recovery and limited effectiveness on low grade minerals; and therefore, the process was regarded cost intensive and inefficient. In addition, the inadequacies of pyrometallurgical extraction of these metals from their complex low grade mineral ores due to its difficulty and large energy requirement in melting the value metals as well as various elemental impurities/gangue minerals have shown great concerns [2]. Despite, the large energy requirement in melting the elemental impurities/gangue minerals as a result of the complexity and low-grade nature of the minerals, successful high temperature processes/applications have been recorded, to adequately/efficiently extract and recover Nb and Ta metals from their minerals. Therefore, on this premise, the study provides insights into recent pyrometallurgical techniques of the extraction and recovery of Nb and Ta, as well as certain associated influencing factors. Thus, this chapter is concisely based on the recent pyrometallurgical approaches relating to the beneficiation, Nb and Ta extraction and recovery from mineral complexes.

1.1 Niobium and tantalum chemistry

Niobium and tantalum were discovered by British scientist, Charles Hatchett in 1801; and Swedish scientist, Anders Ekeberg in 1802, respectively. Niobium (Nb, Z = 41) has very similar geochemical properties, electronic structures and behavior (small radius/atomic size and high charge) with tantalum (Ta, Z = 73) [3, 4] and as such the two metallic elements are regarded as geochemical twins [5–7], and hence similar beneficiation and extraction techniques are usually adopted and employed for both metal elements. Nb and Ta metals are both refractory, transition, weakly acidic, BCC solid metallic elements. A lot of research work and investigations have been reported on the extraction and recovery of Nb and Ta as well as the chemistry involved in extracting both metals. From studies conducted, the Nb and Ta chemistry, hydrolysis, solubility and complexes in aqueous media have been established by various researchers [2, 8–16]. From the reports by certain researchers, it can simply be established that oxygen (O_2^-) ions have often shown very close similarities in ionic radii compared to that of fluorine (F^-) ions. Thus, there can be somewhat easy/simple substitutions occurring between O_2^- and F^- anions in a matrix or complexing compounds [2, 11]. In addition therefore, the hydroxides (OH^-) also possess somewhat similar chemical characteristics/properties as well as in geometry with F^- anions. This similarity is especially and essentially found in their charge and size; and thus, a justification (scientifically), of somewhat easy and feasible substitutions that may occur between their complexes/compounds in the chemical and/or thermodynamic reactions. Therefore, apart from O^{2-}, Ta and Nb are hypothesized also, to be very liable of forming or producing soluble complexes or compounds with certain strong ligands, such as: OH^- and F^- [2, 17, 18]. More so, Nb and Ta are particularly somewhat soluble (in wt. % levels) with alkaline melts or solutions [2, 18, 19], and might even as well attain greater solubility with carbonatite molten solutions [2, 18, 20]. **Figures 1** and **2** depicts the basic description and representation of Nb and Ta metals, respectively, indicating that they are refractory, transition, weakly acidic, BCC solid metallic elements.

1.2 Niobium and tantalum minerals

Nb and Ta primary mineral ores are natural occurring complex oxide minerals, mostly of low grade deposits with chemical composition majorly composed of different content levels of Nb and Ta refractory metals' penta-oxides (Nb_2O_5 and Ta_2O_5), iron (Fe_2O_3), manganese (MnO), tin (SnO_2) and titanium (TiO_2) oxides with other refractory, metallic oxides. Nb/Ta minerals occurring as penta-oxides (Nb_2O_5/Ta_2O_5) are often associated together in similar mineral structures. Over the years, a lot of researchers have consistently and extensively studied the recovery and extraction of Ta and Nb from their primary deposits or mineral ores as well as their secondary sources, applying various mineral processing steps and certain beneficiation/separation methods. More recently, Nb and Ta have been extracted from several primary rock deposits and mineral ores. Howbeit, these primary deposits/mineral ores of

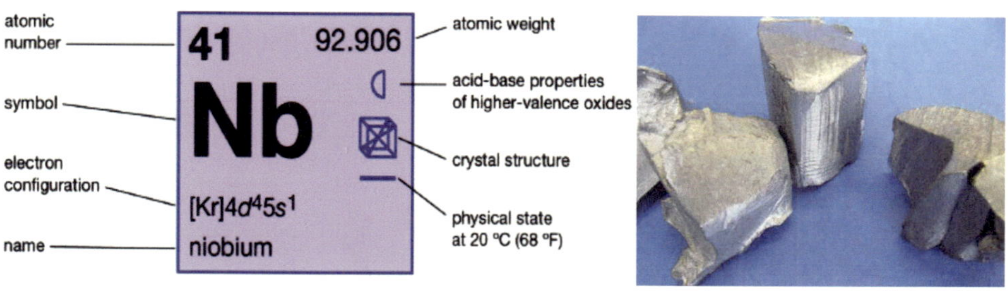

(a) (b)

Figure 1.
(a) Nb element; (b) samples of Nb metal [21–23].

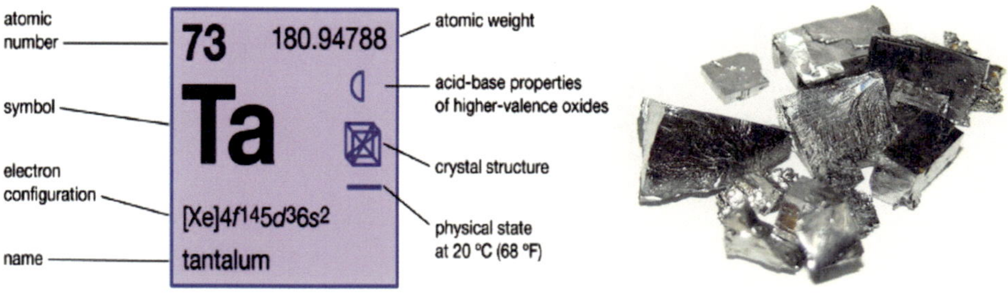

(a) (b)

Figure 2.
(a) Ta element; (b) samples of Ta metal [21, 23].

Ore Mineral	Chemical formula	Locations	Major characteristics/properties			
			Color	S.G	Structure	Deposition
Columbite-tantalite (Coltan)	$(Fe,Mn)(Nb,Ta)_2O_6$	Australia, Brazil, DRC, Ethiopia, Finland, France, Japan, Madagascar, Nigeria, Norway, Russia, USA	Reddish-brown to black	5.2–8.2	Orthorhombic	Peralkalines,Beryl pegmatites (granitic &syenitic), Carbonatites, Peraluminous granites
Pyrochlore	$(Na,Ca)Nb_2O_6(O,OH,F)$	Brazil, Canada, DRC, Norway	Pale yellow to dark black	4.2–5.8	Isometric; Octahedral	Pegmatites, Peralkalines,Carbonatites
Microlite	$(Na,Ca)_2Ta_2O_6(O,OH,F)$	Australia, Brazil, Madagascar, Norway, Sweden, USA, Zimbabwe	Brown to pale yellow	5.5–6.4	Isometric; Cubic isomorphic	Pegmatites, Peralkalines,Carbonatites
Fergusonite	$(REE)(Ba,Nb,Ta)O_4$	Australia, Madagascar, Norway, Rhodesia, Russia, Srilanka, Sweden, USA	Yellow to brown	4.3–5.8	Tetragonal; Monoclinic	Granite pegmatites
Samarskite	$(Fe,Ca,U,Y,Ce)_2(Nb,Ta)_2O_6$	Canada, DRC, India, USA	Black to reddish-brown	4.2–6.2	Orthorhombic	Granitepegmatites
Mossite-Tapiolite	$(Fe,Mn)(Nb,Ta,Ti)_2O_6$	Morocco	Brownish-black to black	7.3–7.9	Tetragonal	Granitepegmatites

Table 1.
Major Nb and Ta mineral resources [2, 11, 23, 25, 26].

Ta and Nb are somewhat scarce in the world and are naturally deposited with a low average crustal abundance of about 2 and 20 mg/kg, respectively. Hence, this in their classification, are referred to as rare metals [15, 16, 23, 24]. **Table 1** displays important Nb-Ta minerals along with their major characters.

1.3 Niobium and tantalum extraction

The demand of Nb and Ta has consistently become somewhat higher than the supply. However, the extraction and recovery of Nb and Ta from their minerals and from other elements contained in the mineral composition, has become very tedious and difficult, involving very complex processes and techniques. Also, the choice selection and application variation of the mineral enrichment, beneficiation and extraction techniques has become somewhat tedious and complicated as they practically depend on the mineralogy of the mineral ore or concentrate, its nature and type, level of purity and impurity, number of safe beneficiation/extraction process steps and most importantly, on the physicochemical properties of the mineral with respect to the chemical composition (the physical and chemical nature/properties of the value and gangue minerals [2, 23]. **Table 2** depicts certain major determinant or influencing factors to be considered when selecting suitable techniques for extraction/recovery of Nb/Ta.

The extraction efficiency of Nb and Ta from mineral deposits greatly depends on the measure or degree of removal of associated impurities present in the crude Nb mineral ore [28]. Hence, for a successful Nb and Ta metal extraction (for both pyrometallurgical and pyro-hydrometallurgical procedures), the impurity content in the oxide mineral ore and the choice of beneficiation/extraction methods and/or extraction agents are key [29]. Thus, a successful beneficiation/separation process application is often rated on the basis of the obtained percentage recovery of the high-purity valuable minerals as well as the impurity content being reduced to a minimum with fewer processing/recovery steps [2, 16]. In addition, low-grade Nb/Ta minerals mined mechanically, such as columbites and tantalites have been reported to contain a lot of impurities and sometimes have less than 0.1% Nb_2O_5/Ta_2O_5 contained in the mineral ore. This increases the difficulty in the mineral processing and subsequent metal recovery. Thus, an efficient ore enrichment beneficiation process route is important to effectively improve the metal extraction and significantly eases up the subsequent downstream pyro- or hydro-metallurgical decomposition and separation processes of the mineral for successful Nb and Ta recoveries. Hence, the need for the minerals to undergo beneficiation/enrichment processes for ore upgrade to an industrially acceptable metallurgical grade of minimum composition of 25% Nb_2O_5 and Ta_2O_5 or 50% in their combine form, which is the required content in the ore for effectiveness/efficiency of subsequent downstream extraction process [30].

Howbeit, due to the complexity and presence of impurities such as: refractory/metallic oxides, rare earths and radioactive elements; Nb_2O_5 and Ta_2O_5 are usually hydrometallurgically extracted from its mineral ore source using hydrofluoric (HF) acid as the decomposition agent before separation and recovery. The traditional method of extraction of Nb and Ta from primary mineral resources using hydrometallurgical procedures namely; employing aqueous dissolution and subsequent solvent extraction (SX), ionic exchange (IX) or other separation/purification processes [31, 32] is well documented in literature. However, some of these process steps are usually employed under harsh aqueous media conditions of very toxic, concentrated, hazardous and corrosive chemicals with several complex separation steps [33].

Factors	Explanation
Technical	The mineralogy, ore type, nature, purity & impurity, scalability & number of beneficiation/extraction process steps
Economical	The cost price of mineral ore type, beneficiation cost & profit, cost per ton of metal produced, fuel availability & cost, quantity/quality of process products, market & energy cost
Environmental	The eco pollution, proper handling & storage of waste/by-products, environment & ISO 14001 guidelines
Quality	The implementation & maintenance of a quality management system in line with ISO 9001 & ISO 17025 standards
Safety	The beneficiation process facility requirements for the safety of workers, safe handling & storage of waste/by-products, maintaining health, safety & OHSAS 18001 standards

Table 2.
Factors considered when selecting Nb and Ta extraction/recovery process routes [16, 23, 27].

This is chiefly due to the insolubility rate of Nb, Ta and other refractory metallic oxides in milder conditions [32, 34]. The chemical and physical similarities of Ta and Nb with other refractory elements make their separation from mineral ores a complex and difficult process [35]. Over the decades, several pyro- and hydro-metallurgical processes of the extraction/recovery of Nb and Ta have been established by various researchers. For instance, Jean Charles Galissard de Marignac developed a hydrometallurgical process in 1866, popularly referred to as the "Marignac process" for Nb and Ta extraction. This very method involved fractional crystallization process in order to separate the metals as potassium oxypentafluoroniobate monohydrate ($K_2[NbOF_5]$ H_2O) and potassium heptafluorotantalate ($K_2[TaF_7]$), respectively. This in turn was also reduced in order to obtain the metals of certain high purity [2, 36], most times adopting the process of electro-winning in fused salts [2, 13]. Howbeit, the establishment of liquid–liquid extraction (LLE) in the 20th century replaced the Marignac method. This process was however developed by the Ames Laboratory and the U.S. Bureau of Mines in 1957. The method utilized the difference in acid solubility of Nb and Ta F^- ions in organic solvents at specific acid levels [2, 10, 37–40].

In recent times, several researchers and investigators enhanced this Nb/Ta extraction method and also, various novel process applications have been studied and developed. The digestion of Ta and Nb mineral particles and subsequent stepwise/selective separation of the metals from their reaction complexes; however considering the recovery and purity level, cost and economic aspects, energy and reagent consumption [2, 41], waste and environment management issues. Among methods adopted by these researchers in order to achieve simultaneous digestion and decomposition of Ta and Nb minerals were chlorination process, alkali fusion method, alkali fusion-acid leaching, alkaline solution dissolution, ammonium fluoride (NH_4F) and ammonium bifluoride (NH_4HF_2) fusion, decomposition with direct H_2SO_4 acid or combined with HF acid, [2, 9, 42–45]. Howbeit, the most successful method employs the initial halogenation and dissolution in HF acid or a combination of HF acid or a fluoride medium like fluoride salts (which forms oxy-fluorides) with mineral acids

[2, 26, 39, 40, 43, 45–48]. This is usually achieved at somewhat high temperatures and concentrations, and under harsh, hazardous and corrosive operational conditions [9, 26, 43, 49, 50] due to the resistance of Nb-Ta mineral ores to chemical/acid attack at mild conditions [32, 39, 40, 48, 51, 52]. Hence, with selective leachants like HF acid, hydrometallurgical techniques was completely adopted and recommended suitable, as a result of its lesser consumption of energy and higher recovery/grade purity of Nb products essentially from complex and low grade mineral ores [52, 53]. However, with stringent regulations on health, safety and environment (HSE) increasing significantly every day, adoption of this hydrometallurgical process has since then, suffered its own share of demerits due to toxicity and harmful nature of HF and other fluoride media [44, 52, 54, 55].

Howbeit, the present limitations involved in the conventional use of HF acid for leaching and dissolution of Nb and Ta minerals ranges from the loss of HF acid through volatilization during the metals' extraction, of about 6–10% Vol. (pressure of 763 mm Hg at room temp), the great amounts of HF acid wastes generated that contains fluoride salts and the serious challenge it poses to personnel and the environment with respect to waste management as well as the safe disposal of such waste fluoride salts. Thus, it can be related that the high corrosiveness, toxicity, volatility, chemical consumption and cost, generation of fluoride-bearing waste water, and other environmental related issues of HF acid, thus contribute to the difficulty and complexity of the extraction/separation process route, as well as the high equipment maintenance/operation cost. Due to these shortcomings as well as the low recovery yield of Ta and Nb, the demand for these metals is somewhat higher than their supply. This is as a result of their criticality and numerous engineering/technological applications, such as: the utilization in the production of rocket missiles, aircraft engines, telephones, solar cells, turbine blades, capacitors, HSLA/stainless steels, oil/gas pipelines, particle accelerators, nuclear reactors, super conductors, refractive index of lenses, heat resistant/cutting tools [2, 23, 56]. Thus, a great deal of research investigation is imperative for enhancing or developing simpler, cleaner, effective and less complicated Ta and Nb extraction/separation process route in order to obtain optimum recoveries and resource utilization. On this premise therefore, several researchers have established the need to develop fluoride-free mineral decomposition media under mild conditions as substitute or alternative extraction procedure for Nb and Ta recovery. Albeit several investigations conducted on Nb and Ta extraction process route from their various mineral resources, research is still on-going in order to completely establish and develop a more efficient process alternative or substitute route for the dissolution/decomposition of the minerals; with very significant efforts ascribed towards mitigating the draw-backs and shortcomings of the conventional extraction/separation of Nb and Ta adopting HF acid as the dissolution medium. As a result of the various challenges encountered in the separation of Ta and Nb from associated impurities in their minerals and from each other, it thus became necessarily important to explore/exploit several extraction and recovery processes. Therefore, it is only imperative that serious attention should also be attributed to the pyrometallurgical as well as the pyro-hydrometallurgical approaches for optimum/efficient Nb and Ta extraction/recovery from their minerals.

2. Pyrometallurgical approaches

Over the years, the pyrometallurgical approach of Nb and Ta extraction has been established. Several authors have reported their investigations on the conventional pyrometallurgical reduction of Nb/Ta penta-oxide to the metal [36]. However, this process often consumed a lot of energy, high reagent consumption, high cost of equipment/equipment maintenance, low Nb/Ta recovery and limited effectiveness on the Nb/Ta low grade minerals; and thus, certain researchers considered the process application inefficient and not cost effective [52]. Habashi [53] reported the inadequacies of the pyrometallurgical extraction of the metals from their complex and low grade minerals due to the extraction difficulty, high temperature and large energy requirement in melting the elemental constituents and gangue minerals. Hence, with selective leachants like hydrofluoric (HF) acid, hydrometallurgical techniques have been completely adopted and recommended suitable, as a result of its lesser consumption of energy and higher recovery/grade purity of Nb products essentially from complex low grade minerals [52, 53]. However, with stringent regulations on health, safety and environment (HSE) protection increasing significantly every day, the adoption of this hydrometallurgical process has since then, suffered its own share of demerits due to the high eco-unfriendliness, toxic and harmful nature of the adopted HF acid and other fluoride extraction media [44, 52, 54, 55], in comparison to the pyrometallurgical techniques. Recently, a lot of research works have been directed towards the pyrometallurgical extraction of Nb and Ta, especially as an adjunct process for optimum and efficient recovery of the metals. More so, despite the inadequacies and shortcomings involved in pyrometallurgical techniques, more especially the employment of elevated temperatures, certain positive effects however can be achieved during high temperature extractive procedures on complex minerals. Thus the following establishments:

- At elevated temperatures, less complex and less expensive reductive agents may be employed.
- There is accelerated rate of reaction at elevated temperatures, hence, leading to high metal recoveries/extractions or productions.
- The rate of reaction doubles after every 10°C increase in temperature; requiring lower activation energy and thus, aids in quicker chemical reactions.
- In employing high temperatures, there is the ability to thermally treat large amount of minerals or ore deposits in compact space giving rise to capital and processing cost effectiveness.
- High temperature extractive procedures such as: fusing can efficiently extract/separate reactive metals in the mineral composition from the mineral deposits.
- As a result of high temperatures, there is the feasibility of shift of reaction as well as the increase in solubility rate of the metals/metallic compounds in aqueous media, even in H_2O.

2.1 Pyrometallurgy and high temperature extraction techniques

Pyrometallurgical procedures on the treatment of Nb and Ta minerals or complexes may involve certain thermal and chemical processes and reactions which may lead to the modification and formation of entirely new mineral phases.

2.1.1 Roasting process

Roasting as a pyrometallurgical procedure involves gas–solid reactions at elevated temperatures with the aim of mineral/metal decomposition, separation and purification. Roasting of mineral complexes or concentrates is often regarded a thermo-chemical process where the chemical conversion of the mineral takes place with oxygen or some other elements or compounds employed at somewhat high temperatures, converting it into another chemical form or state. This is a route or step that has been adopted by several researchers in the processing of certain Nb and Ta complexes or mineral concentrates, usually conducted in order to prepare the mineral for an adjunct hydrometallurgical procedure. For instance, the feasibility of leaching or dissolution process is at an increased measure if the metals in question are in their less stable (oxide) forms and thus more soluble and easy to dissolute. This is usually the situation after the mineral complex has undergone roasting procedure, either on the Nb and Ta complexes or on other mineral impurities/gangue associated in the mineral matrix. Howbeit, before roasting is performed, the minerals often times undergo some physicochemical beneficiation and separation process to some extent, such as froth flotation, gravity, magnetic or electrostatic concentrations. The mineral concentrate may be mixed with certain reagents or materials in order to aid/enhance roasting. Mineral oxides can be somewhat easily reduced to their metallic state or elemental form compared to sulphide minerals. Typically, the roasting of Nb and Ta mineral complexes may often involve certain complex thermal and chemical gas–solid reactions taking place between the mineral/material solids and the furnace atmosphere; and this however consists of several process types.

2.1.2 Roasting types

2.1.2.1 Oxidative roasting

This is the most common type of roasting process industrially practiced and it's more often employed on sulphide complexes or mineral concentrates. In this type of roasting process, the (sulphide) complex is converted to oxides and sulfur is thereby given off as sulfur dioxide gas (SO_2). The complex is however heated to elevated temperatures in the presence of excess oxygen or air in order to completely or partially burn out or replace the impurity elements (which in most cases is usually sulfur) with oxygen. Here, introduced oxygen is supplied and hence replaces the sulfur being burnt off. However, this procedure is usually considered environmentally unfriendly as harmful sulfuric gases are released into the environment. Howbeit, it is essential to note that the sulfur gas released in the form of sulfur dioxide gas (SO_2) may be trapped and essentially utilized in the production of sulfuric acid (H_2SO_4). At the occurrence of complete or almost complete sulfur removal from the sulphide complex, the residue is referred to as dead roast. In this type of roasting, quartz and certain other gangue minerals may act as catalyzers. The general oxidative roasting chemical reaction is represented in Eqs. 1 to 2.

Metal sulphide ⟶ Metal oxide.

$$MS_{(c)} + 3/2\, O_{2(g)} \rightarrow MO_{(c)} + SO_{2(g)} \quad (1)$$

$$2MS_{(s)} + 3O_{2(g)} \rightarrow 2MO_{(s)} + 2SO_{2(g)} \quad (2)$$

2.1.2.2 Chloridizing roasting

This roasting process type involves the transformation or conversion of certain mineral complexes or metallic compounds to chloride forms either by employing reduction or oxidation processes or conditions. This type of roasting can be employed to extract and process certain metals from their chloride state or forms; such as Nb, Ta, Be, Ti, Zr, as well as U and certain rare earth elements (REE). Some overall chloridizing roasting forms is represented by the chemical reactions in Eqs. 3 and 4. The Eq. 3 represents the chlorination process of a sulphide mineral complex which involves an exothermic reaction; while Eq. 4 presents the involvement of an oxide mineral complex with the addition of elemental sulfur serving as a catalyst and facilitating the chemical reaction. More so, some carbonate mineral ores also react almost similarly as the oxide mineral ores after being decomposed or digested under certain elevated temperatures to their oxide state.

$$2NaCl + MS + 2O_2 \rightarrow Na_2SO_4 + MCl_2 \tag{3}$$

$$4NaCl + 2MO + S_2 + 3O_2 \rightarrow 2Na_2SO_4 + 2MCl_2 \tag{4}$$

2.1.2.3 Volatilizing roasting

This roasting type is the volatilization or elimination of the metallic oxides such as ZnO, As_2O_2, Sb_2O_2 and some other oxides from the mineral complex. Volatilizing roasting however involve the careful oxidation of mineral ores at high temperatures in order to remove unwanted elements or impurities in their volatile oxide state or form. In this type of roasting process, the careful control of the amount of oxygen in the roaster is imperative as the excessive oxidation of the mineral complex may produce non-volatile oxide complexes.

2.1.2.4 Magnetic roasting

This type of roasting involves the controlled reduction of certain metallic mineral complexes, converting them into their metallic forms, in order to ease up subsequent beneficiation, separation and processing routes or steps. For instance, a non-magnetic iron oxide, such as hematite (Fe_2O_3) is reduced in a controlled reduction reaction to magnetite (Fe_3O_4), a more magnetic iron form or state, prior to magnetic separation and/or other concentration processes. This method of roasting has been employed for the concentration of certain Nb and Ta mineral complexes by several researchers with the aim of recovering or eliminating associated iron content.

2.1.2.5 Reductive roasting

This type of roasting process involves the partial reduction of an oxide mineral ore, in preparation to the actual smelting reduction process or hydrometallurgical extraction or recovery of the metal from the mineral ore to its metallic form or state. This is regarded as the most adopted and successful form of roasting on Nb and Ta minerals/complexes.

2.2 Calcination process/thermal treatment

Prior to calcination, a precipitation process is usually conducted on the Nb and Ta products/leachates obtained fromthe downstream hydrometallurgical procedures, generating certain intermediates of Nb and Ta as precipitates, and often in the form of pentaoxides or probably dry hydroxides. The structure of the washed/filtered precipitates may contain either two kinds of water. The first kind of water is incorporated within the Nb(V) and Ta(V) hydroxides, and can only be eliminated with calcination employed at somewhat higher temperatures within the range of 700 to 900°C. The other kind of water is known to be intrinsic moisture and may easily be eliminated by only employing a drying process at lower temperature range within 100 to 200°C.Calcination of the processed Nb and Ta complexes or precipitates may result in the yield of high purity Nb and Ta oxides. Calcination process however, also leads to the elimination of certain volatile constituents that were entrained in the precipitates, such as ammonium or fluoride [2, 25]. The calcination/water removal process may be simply presented as in Eqs. 5 and 6.

$$2Nb(OH)_5 + Heat_{(700-900^{o}C)} \rightarrow Nb_2O_5 + 5H_2O \tag{5}$$

$$2Ta(OH)_5 + Heat_{(700-900^{o}C)} \rightarrow Ta_2O_5 + 5H_2O \tag{6}$$

2.3 Thermal production of Nb and Ta metals

The yield or production of Nb and Ta as pure metals is essentially dependent on their forms which they were found in the metal intermediates. For instance, the Nb metal is often achieved by the reduction of its intermediates employing reducing chemical agents like Al, Ca, Mg, H, Na or any suitable electron donors. The most widely practiced/adopted approach is based on the use of Al and is referred to as aluminothermic process, consisting of reactions between the pentaoxides (Nb_2O_5) and the pure Al at certain temperatures >1100°C, as shown in Eq. 7. However, the Ta metal refining from potassium heptafluorotantalate is mostly obtained by its reduction with Na (in Eq. 8). It is recommended however, that in order to achieve high purity Ta metal, both the salt as well as the Na-bearing reducing agent should be of high purity. The chemical reaction is to take place at approximately 1200°C temperature inside the reactor that operates under an inert atmosphere or vacuum, in order to prevent the accumulation of oxygen during the reduction procedure. The chemical reaction that occurs is therefore exothermic and have been reported to release about an energy of about 2985 kJ/kg [25, 57].

$$Nb_2O_5 + \frac{10}{3}Al \rightarrow 2Nb_{(s)} + \frac{5}{3}Al_2O_{3(s)};\ \Delta H = -890 kJ/mol \tag{7}$$

$$K_2TaF_7 + 5Na \rightarrow Ta + 5NaF + 2KF \tag{8}$$

2.4 Influencing factors

The selection of appropriate existing, developing or newly developed novel beneficiation and extraction process routes, decomposition and separation techniques that are suitable for complex Ta and Nb mineral matrices and concentrates is highly dependent on specific factors. Other than the physical and chemical factors or properties, which includes: the mineral type, nature, and chemical composition, there are other factors that may affect the selectivity/choice of the process steps and dictate the process application of extraction and separation process routes, such as: the technical, environmental, economical, safety and quality factors. This may also include the ability/feasibility of obtaining high purity/high yield recoveries, with minimal consumption of energy, number of process steps, by-products and environment pollution. Cost and scalability from laboratory to industrial/commercial scales are also considered major factors [16, 23]. The cost factor however may include the price cost of the mineral type, price cost of the processed materials, fuel cost and availability as well as product quality/quantity [23, 27]. More so, certain determinant factors or parameters may influence the degree of success as well as the decomposition rate of the roasting of Nb and Ta complexes. This includes: temperature, time, physicochemical condition of the mineral complex, chemical reagent type, availability/cost of the reagents or additives, mineral sample to reagent ratio, reagent concentration, mixing ratio, stirring speed, stirring time, equipment or furnace as well as the eco-friendly/unfriendly nature of the chemical reagents employed and the reaction process.

3. Literature overview of the pyro-hydrometallurgical recovery of Nb and Ta

Various pyrometallurgical processes have been adopted by several researchers for the extraction and recovery of Nb and Ta. Howbeit, the high temperature and high thermal energy consumption has poised a major setback. Thus this very process is usually adopted with certain adjunct processes, in order to reduce the temperature applied, thermal energy consumption as well as the cost implications. Moving forward, a lot of researchers have concurred that alkali fusion, an old and earliest established industrial method for mineral decomposition [43–45], and the alkali roasting-leaching decomposition process have been employed on Nb and Ta minerals with certain degree of success. The reaction chemistry, complexes and solubility rate or efficiency of Nb and Ta in alkaline media has thus, been investigated by various researchers [18, 19, 43–45, 47, 58–66]. Alkali roasting technique as a decomposition process has also been investigated by researchers [26]. Few investigations have been reported on the efficacy of alkali reductive roasting and dissolution of Nb/Ta ore minerals or concentrates. Nevertheless, there is still need to adopt mild concentrations/consumption of the alkali reagents with reduced/moderate temperatures. Also, most successful decomposition processes were however established effective to a certain measure on high-grade minerals but not effective on low-grade minerals [2, 4]. However, alkali roast and leach dissolution method employing KOH roasting and H_2O leaching on a low-grade Nb-Ta mineral ore has had great recoveries [44, 47]. $KHSO_4$ fusion as a pyrometallurgical process to improve the subsequent leaching efficiency of a low grade Nb/Ta polymineralized ore has also been adopted [62]. The study reported the feasibility of Ta/Nb decomposition and leach recovery efficiencies of alkali dissolution process on low-grade minerals. Hence, the reductive roast mineral treatment utilizing certain alkali agents was regarded an eco-friendly process and thus, a promising extraction route with moderate leachant and energy consumptions. Howbeit, high roasting temperatures and alkali reagent consumptions were main set-backs.

For instance, Ghambi et al. [56] investigated a novel pyrometallurgical procedure for extracting and purifying columbite and tantalite concentrates (29% Ta_2O_5 and 16% Nb_2O_5). A reductive roasting procedure was employed in the study. The Nb and Ta concentrates were reduced with solid carbon in the form of activated charcoal (carbothermic reduction) and alkali as the reducing atmosphere employing high temperatures ranging from 800 to 950°C. This was reported to aid the reduction of the iron oxides present in the concentrates to metallic iron as well as the subsequent magnetic separation. Suri *et al.* [58] investigated the alkaline decompositions for recovery of Nb and Ta from a cassiterite (Sn) bearing mineral, using Na_2CO_3 and K_2CO_3 salts. The authors reported successful recoveries of Nb and Ta employing alkaline fusion/reduction roasting-acid leaching processes. Phase transformations of pellets of Sn oxides such as SnO and SnO_2, and Na_2CO_3 alkaline salt when roasted under certain carbothermic (CO-CO_2 mixed gas) thermal atmospheres were investigated by Liu *et al.* [66]. The resulting highly pure pellets of Na_2SnO_3 reported by the researchers confirm the favorable reaction chemistry of certain alkaline salts on the pyrochlore mineral group. Nete *et al.* [67] however explained the feasibility of adopting the alkali flux decomposition process as a substitute digestion method. Albeit the low Nb and Ta recoveries of microwave digestion, their investigations displayed process simplicity and also showed that the process allowed smaller amounts of chemical reagents for digestion. The authors also concluded that flux fusion with lithium tetraborate had shorter time for digestion and high recoveries of >90%; however larger amounts of chemical reagents and time-consuming sample preparations were needed for digestion.

As a result of the volatilization and loss of toxic/corrosive HF acid during dissolution process of Nb-Ta minerals, resulting to equipment corrosion and harm to human beings and environment, Yang *et al.* [45] established a novel process decomposition of low-grade Nb/Ta minerals. The researchers employed alkali fusion method, applying caustic soda (sodium hydroxide, NaOH) with minimum alkali consumption of alkali-sample ratio reduced from 3:1 to 1:1. A reduced alkali-sample ratio of 1:1, initial mineral particle size of 75 μm and 650°C temperature for 30mins had significant influence on the Nb/Ta mineral decomposition. The mineral was however converted to $Na(Nb,Ta)O_3$ complexing compounds and not that of $Na_3(Nb,Ta)O_4$, which is usually realized from the conventional alkali fusion method. Results however indicated mineral decomposition efficiencies of 98% cassiterotantalites and 99% pyrochlores in the Nb/Ta ore. Howbeit, $K(Nb,Ta)O_3$ and $Na(Nb,Ta)O_3$ compounds/complexes are somewhat not soluble in most mild solutions. Also, Berhe *et al.* [68] carried out the comparison of the performance and decomposition of the kentichamangano-tantalite ore by both HF-H_2SO_4 leaching and KOH fusion-leaching. The comparative study thus proposed KOH sub-molten salt as a substitute process method for the dissolution of Nb/Ta minerals with the intention of mitigating/eliminating HF acid pollution and associated harm or hazards. Results also explained that the dissolution rate is dependent on the proportions/concentrations of the HF-H_2SO_4 and KOH leaching agents. In addition therefore, higher ratios of HF to H_2SO_4 in the acidic system and higher concentrations of KOH during the alkaline fusion process step led to lesser amounts of leftover residues after the decomposition process thereby resulting to better or increased dissolution rates after H_2O leaching. Thus, the authors however concluded that KOH possess the potential of being a suitable alternative or substitute material to the conventionally utilized HF acid. Similarly, De Oliveira, De Souza and Lopes-Moriyama [69], conducted investigations on alkaline thermal treatments with subsequent acid leaching on a ferro-columbite mineral in order to obtain mixed proportions of Ta and Nb oxides. The study however reported good yield recoveries, employing alkali fusion with

Reference	Mineral deposit	Pyro(hydro) metallurgical technique employed	Characterization analysis	Positive results obtained
De Oliveira et al. [69]	Natural Ferro-columbite	$KHSO_4$ fusion + HCl& HF-H_2SO_4 dissolution	XRF, XRD, BET, SEM, EDXS	Increased grade of 99.6% Nb
Sun et al. [71]	Nb-Ti synthetic ilmeno-rutile	Akali (NaOH) roasting + H_2SO_4 dissolution	XRD, XRPS, LPA, ICP-OES	Increased recovery of 96.68% Nb& Ti; Formation of a new Nb-Ti phase for feasible downstream treatment
Ghambi et al. [56]	Columbite& tantalite concs	Na_2CO_3-activated charcoal assisted alkali reductive roasting +2-stage wet MS + H_2O & $C_2H_2O_4$ dissolution + $NaHSO_4$ roasting	XRPD, SEM, EDXS	Concs can attain a purity of 80% with further refining; May serve as alternative approach/ substitute for HF dissolution, mitigating associated demerits
Purcell et al. [33]	Ferro-columbite	NH_4HF_2 fusion/ leaching	XRF, ICP-OES	Recovery of 96.00% Nb& 91.00% Ta
Nete et al. [35]	HG tantalite	NH_4HF_2 fusion + SX with MIAK	ICP-OES	Recovery of ~100.00% Nb& 98.90% Ta
Wu et al. [72]	LG Nb-Ta ore	H_2SO_4 acid roasting/ leaching	MLA, XRF, AAS	Recovery of 84.00% Nb
Permana et al. [73]	Sn slag	NaOHalkali roasting + water quenching + sieving	XRF, TGA, SEM	Recovery of 0.90% Nb& 0.40% Ta
		NaOHalkali roasting + HF/HCl acid leaching		Recovery of 2.09% Nb&2.01% Ta
Permana et al. [74]	Sn slag	NaOHalkali roasting + NaOH quenching	XRF, XRD	Recovery of 0.20% Nb& 0.14% Ta
		NaOHalkali roasting + HCl leaching		Recovery of 3.57% Nb& 3.75% Ta
Kitungwa et al. [75]	Columbite	NH_4HF_2-KOH fusion system	XRF, OPM, ICP-OES	Recovery of 95.00% Nb& 94.00% Ta
Habinshuti et al. [76]	Ferro-columbite	KOH-assisted alkali roasting + water-based leaching	XRF, SEM, EDS, ICP-OES	Recovery of 97.00% Nb& 85.00% Ta
Yang et al. [45]	LG Nb-Ta ore	NaOH fusion + water leaching	XRD, SEM, ICP-OES	Recovery of 99.00% Nb& 98.00% Ta
Wang et al. [44]	LG Nb-Ta ore	KOH alkali roasting + water leaching	XRD, ICP-OES	Recovery of 94.70% Nb& 93.60% Ta
Shikika et al. [77]	Coltan ore	Alkali (KOH) roasting + water leaching	XRD, XRF, SEM, EDS, PIXE, ICP-AES	Decomposition of Nb& Ta without the use of HF; Obtaining PLS suitable for downstream SX

Reference	Mineral deposit	Pyro(hydro) metallurgical technique employed	Characterization analysis	Positive results obtained
Berhe et al. [68]	LG Mangano-tantalite	KOH fusion + water leaching	XRF, XRD, FT-IR	Recovery of 0.35% Nb& 14.20% Ta
Berhe et al. [78]	HG tantalite	KOH fusion + water leaching	EDXRF, XRD, FT-IR, ICP-OES	Recovery of 94.73% Nb& 75.80% Ta
		KOH fusion + water leaching + SX with EMIC/$AlCl_3$		Recovery of 99.84% Nb& 90.81% Ta
Nete et al. [39]	Pure (Nb_2O_5 and Ta_2O_5) pentaoxides	NH_4HF_2 fusion + PPT with PPDA	ICP-OES	Recovery of 23.00% Nb& 73.00% Ta
		NH_4HF_2 fusion + single step SX with MIBK		Recovery of 2.00% Nb& 80.00% Ta
		NH_4HF_2 fusion + double step SX with MIBK		Recovery of ~ 100.00% Ta
		NH_4HF_2 fusion + IX with strong amberlite resin		Recovery of 91.69% Nb& 73.39% Ta
		NH_4HF_2 fusion + IX with weak dowex marathon resin		Recovery of 96.05% Nb& 52.34% Ta
Nete et al. [67]	Nb-Ta tantalite	Microwave assisted H_2SO_4 leaching	XRF, XRD, ICP-OES	Recovery of 90.25% Nb& 88.90% Ta
		$Li_2B_4O_7$ fusion		Recovery of 98.50% Nb& ~ 100.00% Ta

Key: MLA = Mineral liberation analyzer;LG = Low grade; HG = High grade; SX = Solvent extraction; IX = Ionic exchange; MIBK = Methyl isobutyl ketone; MIAK = Methyl isoamyl ketone; EMIC = 1-ethyl-3-methyl imidazolium chloride; PPDA = ρ-phenylenediamine; LPA = Laser particle analyzer; PPT = Precipitation; EDXRF = Energy dispersive X-ray fluorescence; XRPS = X-ray photoelectron spectroscopy; XRPD = X-ray power diffractometry; Concs = Concentrates; SEM = Scanning electron microscopy; EDXS = Energy dispersive X-ray spectroscopy; OPM = Optical microscopy; AAS = Atomic absorption spectrophotometry; BET = Brunauer–Emmett–Teller; ICP-OES/AES = Inductively coupled plasma-optical/atomic emission spectroscopy; FT-IR = Fourier transform-infrared spectroscopy; TGA = Thermogravimetric analysis.

Table 3.
Nb-Ta pyro(hydro)metallurgical extraction procedures adopted by various researchers.

potassium bisulphate ($KHSO_4$) and subsequent HCl acid leaching. More so, mechanical pre-mixing process has been established on primary minerals and reported to aid proper distribution of the chemical reagents on the mineral particles, thus, enhancing pyrometallurgical decomposition [70]. **Table 3** therefore summaries the major pyro(hydro)metallurgical extraction methods recently adopted by researchers.

4. Conclusion

The separation and recovery of Nb/Ta metallic elements from mineral deposits has become somewhat challenging, involving several complex/complicated extraction process routes from the rest of the gangues/impurity elements composed in the mineral ore or deposit and from the value metal elements themselves. This can be chiefly attributed to the presence of numerous impurities such as: metal oxides, refractory, rare earth (REE) and radioactive elements, and hence the complexity of the metal extraction process. In that regard, Nb_2O_5 and Ta_2O_5haveconventionally been extracted hydrometallurgically, using HF acid as the dissolution agent, or a combination of HF and H_2SO_4 mineral acid dissolutions. Howbeit, the high volatility, corrosive and toxic nature, high chemical/reagent consumption, eco unfriendliness and waste generation of fluorides have contributed to Nb/Ta extraction complexity, high equipment maintenance and high operational cost. This, as a result has increased the extraction difficulty and low yield recovery/purity of the metals, coupled with their numerous industrial applications in engineering and technology. This study therefore has provided insights into recent adopted pyro- or pyrohydro-metallurgical Nb/Ta extraction process routes as well as its associated factors. Certain pyrometallurgical techniques, such as reductive roasting, carbothermic reductions, alkali roasting and alkali fusions, etc. have shown great success in Nb/Ta extraction. Regardless of the drawbacks encountered, such as elevated temperature and high energy/reagent consumption, the adoption of high temperature procedures is imperative as it plays key roles in the decomposition of these value metals, in preparation for subsequent downstream measures. Thus, the employment and development of pyrometallurgical extraction routes for efficient Nb/Ta recovery is therefore encouraged with certain process advances/improvements, proper process optimization, as well as adoption of adjunct techniques so as to mitigate/curtail such limitations.

Acknowledgements

We sincerely acknowledge the Faculty of Engineering and the Built Environment, Tshwane University of Technology, Pretoria, South Africa. Most especially, our special recognition and hearty appreciation goes to Adeline C. Nzeh and also to Dr. Swithin N. Nzeh (G.P, United Kingdom; MBBS, MRCGP, Dip. Paed, RCPathME) for their significant contributions towards the success and actualization of the present study.

Author details

Nnaemeka Stanislaus Nzeh[1*], Maite Mokgalaka[1], Nthabiseng Maila[1], Patricia Popoola[1], Daniel Okanigbe[1], Abraham Adeleke[1,2] and Samson Adeosun[3]

1 Department of Chemical, Metallurgical and Materials Engineering, Tshwane University of Technology, Pretoria, South Africa

2 Department of Materials Science and Engineering, ObafemiAwolowo University, Ile-Ife, Nigeria

3 Department of Metallurgical and Materials Engineering, University of Lagos, Akoka, Nigeria

*Address all correspondence to: nstannzeh@gmail.com

References

[1] Hiskey B. Metallurgy Survey in Kirk-Othmer Encyclopedia of Chemical Technology. Weinheim: Wiley-VCH; 2000

[2] Nzeh NS, Adeosun S, Popoola AP, Adeleke A, Okanigbe D. Process applications and challenges in mineral beneficiation and recovery of niobium from ore deposits – A review. Mineral Processing and Extractive Metallurgy Review. 2022;**43**(7):833-864. DOI: 10.1080/08827508.2021.1964965

[3] Htwe HH, Lwin KT. Study on extraction of niobium oxide from Columbite-tantalite concentrate. World Academy of Science, Engineering and Technology, International Journal of Chemical and Molecular Engineering. 2008;**2**(10):223-225

[4] Gao WC, Wen JK, Wu B, Shang H. Extraction of niobium, yttrium, and cerium from a low-grade niobium-bearing ore by roasting (NH4)2SO4-Na2SO4-H2SO4 system. Rare Metals. 2014;**33**(6):754-760

[5] Münker C, Pfänder JA, Weyer S, Büchl A, Kleine T, Mezger K. Evolution of planetary cores & earth-moon system from Nb/Ta systematics. Science. 2003;**301**:84-87

[6] Burnham AD, Berry AJ, Wood BJ, Cibin G. The oxidation states of niobium and tantalum in mantle melts. Chemical Geology. 2012;**330-331**:228-232

[7] Timofeev A, Migdisov AA, Williams-Jones AE. An experimental study of the solubility and speciation of tantalum in fluoride-bearing aqueous solutions at elevated temperature. Geochimica et Cosmochimica Acta. 2017;**197**:294-304

[8] Fairbrother F. The Chemistry of Niobium and Tantalum. Amsterdam; New York: Elsevier Pub. Co; 1967

[9] Gupta CK, Suri AK. Extractive Metallurgy of Niobium. London: CRC Publishing Press, Inc.; 1994. pp. 98-127

[10] Agulyansky A. The Chemistry of Tantalum and Niobium Fluoride Compounds. First ed. Oxford, UK: Elsevier; 2004

[11] Nete M. Dissolution and Analytical Characterization of Tantalite Ore, Niobium Metal and Other Niobium Compounds [Thesis]. Bloemfontein: University of the Free State; 2009

[12] Peiffert C, Nguyen-Trung C, Palmer DA, Laval JP, Giffaut E. Solubility of B-Nb2O5 and the hydrolysis of niobium (V) in aqueous solution as a function of temperature and ionic strength. Journal of Solution Chemistry. 2010;**39**:197-218

[13] Lantelme F, Groult H. Niobium and titanium electrowinning in fused salts. Mineral Processing and Extractive Metallurgy. 2010;**119**(2):82-87

[14] Nikishina EE, Lebedeva EN, Drobot DV. Niobium- and tantalum-containing oxide materials: Synthesis, properties and application. Inorganic Materials. 2012;**48**(13):1243-1260

[15] Koko F. Niobium and Tantalum Beneficiation from Tantalite Ore [Thesis]. Bloemfontein: University of the Free State; 2013. pp. 1-26

[16] Nete M. Separation and Purification of Niobium and Tantalum from Synthetic and Natural Compounds [Thesis]. Bloemfontein: University of the Free State; 2013. pp. 1-36

[17] Wood SA. The aqueous geochemistry of zirconium, hafnium, niobium and tantalum. In: Linnen RL, Samson IM, editors. Rare Element Geochemistry and Mineral Deposits. Vol. 17. Canada: Geological Association of Canada; 2005. pp. 217-268

[18] Linnen RL, Trueman DL, Burt R. Chapter 15, tantalum and niobium. In: Critical Metals Handbook. First ed. UK: John Wiley & Sons Ltd.; 2014. p. 361

[19] LINNEN RL, CUNEY M. Granite-related rare-element deposits and experimental constraints on Ta-Nb-W-Sn-Zr-Hf mineralization. In: Linnen RL, Samson IM, editors. Rare Element Geochemistry and Mineral Deposits. Vol. 17. Canada: Short Course: Geological Association of Canada; 2005. pp. 45-68

[20] Mitchell RH. Mineralogical and experimental constraints on the origins of niobium mineralization in carbonatites. In: Linnen RL, Samson IM, editors. Rare Element Geochemistry and Mineral Deposits. Vol. 17. Canada: Geological Association of Canada; 2005. pp. 201-215

[21] Encyclopaedia Britannica. Niobium chemical element [Online]. 2020. Available from: https://www.britannica.com/science/niobium. [Accessed: August 27, 2020]

[22] HAINES & MAASSEN. Niobium. Metallhandlesgesellschaft [Online]. 2020. Available from: https://www.haines-maassen.com/en/nb. [Accessed: June 29, 2020]

[23] Nzeh NS, Popoola API, Adeleke AA, Adeosun SO. Factors and challenges in the recovery of niobium and tantalum from mineral deposits, recommendations for future development – A review. Materials Today: Proceedings, Elsevier. 2022;**65**:2184-2191. DOI: org/10.1016/j.matpr.2022.06.034

[24] Deblonde GJP, Chagnes A, Weigel V, Cote G. Direct precipitation of niobium and tantalum from alkaline solutions using calcium-bearing reagents. Hydrometallurgy. 2016;**165**:345-350

[25] Shikika A, Sethurajan M, Muvundja F, Mugumaoderha MC, Gaydardzhiev ST. A review on extractive metallurgy of tantalum and niobium. Hydrometallurgy. 2020;**198**:1-13

[26] Sanchez-Segado S, Ruzaidi AF, Zhang Y, Jha A. Characterization of physico-chemical changes during the alkali roasting of niobium and tantalum oxides. In: Drying, Roasting and Calcining of Minerals. USA: The Minerals, Metals & Materials Society; 2015. pp. 51-58

[27] Onyedika GO, Achusim-Udenko AC, Nwoko CIA, Ogwuegu MOC. Chemistry, processes and problems of complex ores utilization: Hydrometallurgical options. International Journal of Chemical Sciences. 2012;**10**(1):112-130

[28] El-Husaini OM, El-Hazek MN. Removal of radioactive elements from niobium and tantalum ores. The European Journal of Mineral Processing and Environmental Protection. 2005;**5**(1):7-16

[29] Zhang Y, Du M, Liu B, Su Z, Li G, Jiang T. Separation and recovery of iron and manganese from high-iron manganese oxide ores by reduction roasting and magnetic separation technique. Separation Science and Technology. 2017;**52**(7):1321-1332

[30] Adetunji AR, Siyanbola WO, Funtua II, Olusunle SOO, Afonja AA, Adewoye OO. Assessment of beneficiation routes of tantalite ores

from key locations in Nigeria. Journal of Minerals & Materials Characterization & Engineering. 2005;**4**(2):85-93

[31] Ungerer MJ, Van Der Westhuizen DJ, Lachmann G, Krieg HM. Comparison of extractants for the separation of TaF5 and NbF5 in different acidic media. Hydrometallurgy. 2014;**144-145**:195-206

[32] Nguyen TH, Lee MS. A review on the separation of niobium and tantalum by solvent extraction. Mineral Processing and Extractive Metallurgy Review. 2018;**40**(4):265-277

[33] Purcell W, Potgieter H, Nete M, Mnculwane H. Possible methodology for niobium, tantalum and scandium separation in ferrocolumbite. In: Minerals Engineering. Vol. 119. Oxford, UK: Elsevier; 2018. pp. 57-66

[34] Deblonde GJP, Weigel V, Bellier Q, Houdard R, Delvallée F, Bélair S, et al. Selective recovery of niobium and tantalum from low-grade concentrates using a simple and fluoride-free process. Separation and Purification Technology. 2016;**162**:180-187

[35] Nete M, Purcell W, Nel JT. Separation and isolation of tantalum and niobium from tantalite using solvent extraction and ion exchange. Hydrometallurgy. 2014;**149**:31-40

[36] Awasthi A, Bhatt YJ, Krishnamurthy N, Ueda Y, Garg SP. The reduction of niobium and tantalum pentoxides by silicon in vacuum. Journal of Alloys and Compounds. 2001;**315**:187-192

[37] Ayanda OS, Adekola FA. A review of niobium-tantalum separation in hydrometallurgy. Journal of Minerals & Materials Characterization & Engineering. 2011;**10**(3):245-256

[38] TIC. Tantalum-niobium international study center [Online]. 2020. Available from: https://www.tanb.org/about-niobium/niobium-history. [Accessed: June 28, 2020]

[39] Nete M, Purcell W, Nel JT. Hydrometallurgical separation of niobium and tantalum: A fundamental approach. JOM. 2016;**68**(2):556-566

[40] Nete M, Purcell W, Nel JT. Separation of niobium and tantalum pentafluoride by selective precipitation using ρ-phenylenediamine. JOM. 2016;**68**(11):2817-2823

[41] Ribagnac P, Deblonde GJP, Blancher SB, Lengagne L, Donati L, Malimba C, et al. Leaching of niobium- and REE-bearing iron ores: Significant reduction of H2SO4 consumption using SO2 and activated carbon. In: Separation and Purification Technology. Vol. 189. Oxford, UK: Elsevier; 2017. pp. 1-10

[42] El-Hussaini OM, El-Hakam Mahdy MA. Sulfuric acid leaching of KabAmiri niobium-tantalum bearing minerals, central Eastern Desert. Egypt. Hydrometallurgy. 2002;**64**:219-229

[43] Zhou H, Zheng S, Zhang Y. Leaching of a low-grade niobium-tantalum ore by highly concentrated caustic potash solution. Hydrometallurgy. 2005;**80**:83-89

[44] Wang X, Zheng S, Xu H, Zhang Y. Leaching of niobium and tantalum from a low-grade ore using a KOH roast-water leach system. Hydrometallurgy. 2009;**98**:219-223

[45] Yang X, Wang X, Wei C, Zheng S, Zhang Y. Decomposition behaviors of tantalum-niobium ore in sodium hydroxide system. Advanced Materials Research. 2012;**361-363**:644-649

[46] El-Hussaini OM, El-Hakam Mahdy MA. Extraction of niobium and tantalum from nitrate and sulfate media by using MIBK. Mineral Processing and Extractive Metallurgy Review. 2002;**22**:633-650

[47] Wang X, Zheng S, Xu H, Zhang Y. Dissolution behaviors of Ta2O5, Nb2O5 and their mixture in KOH and H2O system. Transactions of Nonferrous Metals Society of China. 2010;**20**:2006-2011

[48] Yang X, Wang X, Wei C, Zheng S, Sun Q, Wang D. Extraction kinetics of tantalum by MIBK from pulp using Lewis cell. Hydrometallurgy. 2013;**131-132**:34-39

[49] He J, Zhang Z, Xu Z. Hydrometallurgical extraction of tantalum and niobium in China. Tantalum-Niobium International Study Centre Bulletin. 1998;**93**:1-6

[50] Welham NJ. Enhanced dissolution of tantalite/columbite following milling. International Journal of Mineral Processing. 2001;**61**:145-154

[51] Nikolaev AI, Maiorov VG. New approaches to niobium and tantalum extraction technology. In: Doklady Chemistry. Vol. 415, Part 1. UK: Springer; 2007. pp. 167-169

[52] Zhang B, Liu C, Li C, Jiang M. Separation and recovery of valuable metals from low-grade REE-Nb-Fe ore. International Journal of Mineral Processing. 2016;**150**:16-23

[53] Habashi F. Metallurgie extractive, chapter 2. In: Textbook of Hydrometallurgy. Second ed. Quebec, Canada: Métallurgie Extractive Québec; 1999. pp. 18-19

[54] Liu ZQ, Chen HJ, Liang ZF, Feng YQ. Method for Decomposing Ore Concentrate Containing Tantalum, Niobium and Rare-Earth Element. China: China Patent; 2008. CN101440429

[55] Zhu Z, Cheng CY. Solvent extraction technology for the separation and purification of niobium and tantalum: A review. Hydrometallurgy. 2011;**107**:1-12

[56] Ghambi S, Sanchez-Segado S, Chipakwe V, Jha A. An investigation on hydrofluoric (HF) acid-free extraction for niobium oxide (Nb2O5) and tantalum oxide (Ta2O5) from columbite/tantalite concentrates using alkali reductive roasting. Minerals Engineering. 2021;**173**:1-11

[57] Mirji KV, Sheela Saibaba N. Technological challenges in extractive metallurgy and refining of Nb, Ta and preparation of their compounds & alloys. Materials Today: Proceedings. 2016;**3**(9):3151-3161

[58] Suri AK, Subramanian C, Gupta CK. Investigations on flowsheet development for the treatment of niobium-tantalum bearing cassiterite. Mineral Processing and Extractive Metallurgy Review. 1992;**9**(1-4):293-303

[59] Zhou H, Zheng S, Zhang Y. Leaching of a low-grade refractory tantalum-niobium ore by KOH sub-molten salt. The Chinese Journal of Process Engineering. 2003;**5**:1-10

[60] Zhou H, Yi D, Zhang Y, Zheng S. The dissolution behavior of Nb2O5, Ta2O5 and their mixture in highly concentrated KOH solution. Hydrometallurgy. 2005;**80**:126-131

[61] Zhou H, Zheng S, Zhang Y, Yi D. A kinetic study of the leaching of a low-grade niobium-tantalum ore by concentrated KOH solution. Hydrometallurgy. 2005;**80**:170-178

[62] El-Hazek MN, Amer TE, Issa RM, Abu El-Azm MG, Omar SA, El-Hady SM. Characterization and breakdown of south Gabal El-A'urfpolymineralized ore material. Eurasian Chemistry Technological Journal. 2009;**11**:149

[63] Yang X, Wang X, Wei C, Zheng S, Sun Q. Decomposition of niobium ore by sodium hydroxide fusion method. The Minerals, Metals & Materials Society and ASM International. 2012;**44B**:45-52

[64] Deblonde GJP, Chagnes A, Bélair S, Cote G. Solubility of niobium (V) and tantalum (V) under mild alkaline conditions. Hydrometallurgy. 2015;**156**:99-106

[65] Deblonde GJP, Bengio D, Beltrami D, Bélair S, Cote G, Chagnes A. A fluoride-free liquid-liquid extraction process for the recovery & separation of niobium & tantalum from alkaline leach solutions. Separation & Purification Technology. 2019;**215**:634-643

[66] Liu B, Zhang Y, Su Z, Li G, Jiang T. Phase evolution of tin oxides roasted under CO-CO2 atmospheres in the presence of Na2CO3. Mineral Processing and Extractive Metallurgy Review. 2016;**37**(4):264-273

[67] Nete M, Purcell W, Snyders E, Nel JT, Beukes G. Characterization and alternative dissolution of tantalite mineral samples from Mozambique. The Journal of the Southern African Institute of Mining and Metallurgy. 2012;**112**:1079-1086

[68] Berhe GG, Velazquez DRA, Bogale T, Abubeker Y, Girma W. Decomposition of the Kentichamangano-tantalite ore by HF/H2SO4 and KOH fusion. Physicochemical Problems of Mineral Processing. 2018;**54**(2):406-414

[69] De Oliveira TF, de Souza CP, Lopes-Moriyama AL. Acid leaching and thermal treatments in the obtaining of mixed oxides of Nb and Ta from ferrocolumbite. Minerals Engineering. 2020;**147**:1-7

[70] Liu H, Xiang N, Sheng X, Zhai Y, Han C. Decrease of material burden in a novel alkali-saving reduction treatment process of nickel slag based on NaOH roasting. In: JOM, Recycling Silicon and Silicon Compounds. USA: The Minerals, Metals & Materials Society; 2019. pp. 1-11

[71] Sun L, Yu H, Meng F, Qi T, Wang L, Peng Y. Recovery of niobium and titanium from ilmenorutile by NaOH roasting-H2SO4 leaching process. Journal of Materials Research and Technology. 2021;**15**:2575-2583

[72] Wu B, Shang H, Wen J-K. Sulfuric acid leaching of low-grade refractory tantalum-niobium and associated rare earths minerals in Panxi area of China. Rare Metals. 2015;**34**(3):202-206

[73] Permana S, Soedarsono JW, Rustandi A, Maksum A. Other oxides pre-removed from Bangka tin slag to produce a high grade tantalum and niobium oxides concentrate. MOIME 2016: IOP Conference Series: Materials Science and Engineering. 2016;**131**:1-9

[74] Permana S, Majid RA, Soedarsono JW. Hydrochloric acid leaching approaches for tantalum and niobium pentoxides enrichment in 1st tin slag from Bangka Island. International Conference on Advanced Materials and Technology, IOP Conference Series: Materials Science and Engineering. 2020;**924**:1-11

[75] Kitungwa BK, Yuma PM, Kateule CM, Kyona CW, Wakenge IB. Optimization of the water leaching

parameters of tantalum and niobium contained in Manono ore in the Democratic Republic of Congo by fusion in the NH_4HF_2-KOH system. IOSR Journal of Applied Chemistry. 2020;**13**(5):1-6

[76] Habinshuti JB, Munganyinka JP, Adetunji AR, Mishra B, Tanvar H, Mukiza J, et al. Caustic potash assisted roasting of the Nigerian ferro-columbite concentrate and guanidine carbonate-induced precipitation: A novel technique for extraction of Nb-Ta mixed-oxides. Results in Engineering. 2022;**14**:1-12

[77] Shikika A, Muvundja F, Mugumaoderha MC, Gaydardzhiev ST. Extraction of Nb and Ta from a coltan ore from south Kivu in the DRC by alkaline roasting – Thermodynamic and kinetic aspects. Minerals Engineering. 2021;**163**:1-11

[78] Berhe GG, DRA V, Abubeker Y, Girma W, Bogale T, Sisay CM. Green extraction of niobium and tantalum from Kenticha tantalite ore using 1-ethyl-3-methyl imidazolium chloride ionic liquid. Journal of Mining & Environment. 2018B;**9**(4):785-794

Chapter 3

Alternative Extraction Systems for Precious Metals Recovery: Aqueous Biphasic Systems, Ionic Liquids, Deep Eutectic Solvents

Olga Mokhodoeva

Abstract

The current trend in the development of separation methodologies implies their evolution in an environmentally friendly perspective, more precisely, the transition to techniques, materials, and solvents that could be qualified as greener alternatives to conventional ones. The green extraction systems can be attributed to aqueous biphasic systems, ionic liquids, and deep eutectic solvents, which have been widely used recently for various analytical, synthetic, and industrial tasks. In this chapter, the features of the listed systems are discussed in relation to the extraction of precious metals, mainly platinum, palladium, and gold; the examples of the alternative extraction systems for separation and preconcentration of precious metals are reviewed.

Keywords: extraction, precious metals, platinum group metals, aqueous biphasic systems, ionic liquids, deep eutectic solvents, recycling, green chemistry

1. Introduction

Platinum, palladium along with their satellites (platinum group metals, PGMs), and gold are essential and often irreplaceable in many areas of science and industry, including rapidly developing fields of electronics, automobile and engineering industry, medicine, and so on. Going forward, there is a high potential for new PGMs' use in energy transition applications [1–3]. The limited natural reserves and the steadily growing demand for precious metals dictate an improvement of technologies for processing the primary raw materials and especially secondary resources, namely spent automotive catalysts and electronic wastes [4–7]. At the same time, selective recovery of precious metals is also an important analytical problem, because it represents a crucial step in their accurate determination. This problem is relevant for searching and evaluation of new deposits and study of alternative sources, as well as for obtaining data on monitoring PGMs in the environment and biological fluids [8].

Various methods for processing platinum-containing materials and recovery of precious metals based on co-precipitation [9, 10], solvent extraction [11], sorption [12], electrodeposition [13], molecular recognition [14], and so forth are explored.

IntechOpen

The vector of progress of the mentioned methods, coinciding with that for chemical and process industries as a whole, is aimed at finding tools that align with the principles of green and white chemistry and the interests of sustainable development [15–17]. From this standpoint, utilization of alternative solvents and extraction systems of new generation is of great scientific and practical importance. As such systems, aqueous biphasic systems, supramolecular solvents, supercritical fluids, and the so-called designer compounds—ionic liquids and deep eutectic solvents—are currently considered. Some of them are addressed in this chapter relating to precious metals extraction for both technological and analytical applications.

2. Aqueous biphasic systems (aqueous two-phase systems)

Phase-forming components of aqueous biphasic systems (ABSs) are water-soluble, nontoxic, biocompatible, available, and produced in large quantities. Recently, along with traditional polymer-salt and polymer-polymer systems, ABSs produced on the basis of non-polymeric compounds, such as organic salts, in particular ionic liquids, hydrophilic solvents (acetonitrile, alcohols), and so on, have been proposed [18, 19].

Technologies for producing and purifying enzymes and other biomolecules using ABSs based on biodegradable polymers have been developed and successfully introduced into industry [20–22]. Studies on the extraction of metals, mainly nonferrous, radioactive, transplutonium, were first described in the 80s of the last century [23].

The data on extraction of precious metals in ABSs are limited to only a few publications (see **Table 1**). One of the first works by Bulgariu [24] described the extraction of gold(III) using an ABS prepared by mixing polyethylene glycol PEG-1500 and $(NH_4)_2SO_4$ aqueous solutions, including in the presence of chloride ion. It has been shown that Au(III) is almost quantitatively (> 98%) extracted into the polymer-rich phase at pH $\leq$ 3.0 and the concentration of chloride ion >0.08 mol L^{-1}. The developed method of gold extraction was tested on the example of electronic wastes containing Cu(II), Co(II), Ni(II), Zn(II), Fe(III), and Pb(II) ions.

Metals	ABSs composition	Extraction conditions	Reference
Au(III)	PEG-1500 – $(NH_4)_2SO_4$	[Au] = 10^{-2} M, pH $\leq$ 3, [Cl^-] > 0.08 M	[24]
Pd(II)	PEG-1500 – $(NH_4)_2SO_4$	[Pd] = 0.02–0.2 mg mL^{-1}, [H^+] = 0.02–0.2 M, [Cl^-] = 0.25·10^{-3}–4 M	[25]
Pd(II)	PEG-1500 – Na_2SO_4	[Pd] = 0.2–1 mM, 0.1 M HCl/(0.05 M H_2SO_4 + 0.1 M NaCl)	[26]
Au(III)	L64 – Li_2SO_4	[Au] = 0.8–21 mg kg^{-1}, (HCl + HNO_3)	[27]
Au(III)	PEG-6000 – $[C_6MIm][C_{12}SO_3]$	[Au] = 0.48 g L^{-1}, pH = 1.13–1.90	[28]
Pd(II)	PEG-2000 – K_2HPO_4 $[C_nMIm]Br$, n = 4,6,8	[Pd] = 200–1200 mg L^{-1}, pH $\leq$ 1	[29]
Pd(II), Pt(IV)	PEG-1500 – $(NH_4)_2SO_4$	[Pd] = 0.7–1.5 g L^{-1}, [Pt] = 0.1–0.7 g L^{-1}, (0.1 M HCl + 10–90 g L^{-1} NaCl), RCC*	[30]

**RCC: rotating coiled column.*

Table 1.
The examples of ABSs application for extraction of precious metals.

Simonova and coauthors studied the extraction of palladium(II) chloride complexes in an ABS (PEG-1500, PEG-115) – NaCl – $(NH_4)_2SO_4$ – H_2O [25]. The experimental results have shown that palladium is extracted into the organic phase by the solvation mechanism as a compound $H_2[PdCl_4]{\cdot}yPEG{\cdot}nH_2O$. Hyphenated methods of spectrophotometric determination of palladium(II) and iridium(IV) and their separation from rhodium(III) and ruthenium(III) have been developed [31].

The effect of acid and chloride ion concentrations on phase equilibria of an ABS PEG-1500 – Na_2SO_4 – H_2O and the partition of palladium were investigated by Milevskiy et al. [26]. The maximum distribution coefficients of palladium were achieved under extraction in ABSs PEG-1500 – Na_2SO_4 – 0.1 M HCl and PEG-1500 – Na_2SO_4 – (0.05 M H_2SO_4 + 0.1 M NaCl).

A method of gold extraction from scrap central processing units with an ABS based on L64 triblock copolymer, lithium sulfate, and matrix ions of the leachate solutions has been proposed [27]. Gold is quantitatively extracted in the macromolecular-rich top phase without the use of any auxiliary extractants; copper is extracted as a by-product.

Extraction of gold(III) from acidic solutions in an ABS based on PEG-6000 and imidazolium ionic liquid was carried out [28]. For this purpose, a new ionic liquid, 1-hexyl-3-methylimidazolium dodecyl sulfonate ($[C_6mim][C_{12}SO_3]$) was synthesized. Under optimal conditions, the degree of gold extraction was 97%.

Tang et al. [29] studied the extraction behavior of palladium(II) from hydrochloric acid solution using a typical polymer-salt ABS with an ionic liquid as a functional additive. It was shown that extraction system based on PEG-2000 and K_2HPO_4 with imidazolium ionic liquid allowed to recover 96–99% of Pd(II) from acidic medium.

The method of extraction and separation of palladium(II) and platinum(IV) in an ABS based on PEG-1500 and ammonium sulfate from model technological solutions under dynamic conditions was developed [30]. In this case, the traditional ABS was firstly applied for extraction in a rotating coiled column (an analog of centrifugal extractor), where a polymer-rich phase is retained as a stationary phase without any solid support. A multistage extraction could be realized in the system according to the countercurrent chromatography theory [30].

3. Ionic liquids

A large number of reviews and experimental original works are devoted to ionic liquids (ILs). With ever increasing interest to fundamental and applied chemistry of ILs, the field shows no signs of slowing down. These compounds, represented as ionically bonded organic cation and organic or inorganic anion, are being defined as molten salts with melting points below 100°C. Due to their low volatility, thermal stability, and high solubility, ILs are considered to be a safer alternative to traditional molecular organic solvents [32, 33]. At the same time, compliance with the requirements for "green" solvents, namely, nontoxicity, stability, renewability, atomic efficiency, and so on, is not always feasible in the case of ILs and is determined by their composition [34, 35].

In the last two decades, studies on the use of ILs in respect to the problem of PGMs and gold separation have been actively carried out. Various ILs classified by cationic groups—ammonium, imidazolium, pyridinium, phosphonium, guanidium, and betaine—are described as exhibiting affinity and selectivity to precious metals. Examples of extraction systems based on ILs for the extraction of precious metals are quite numerous. Several comprehensive reviews on the subject have been recently published by Lee [36], Firmansyah [37], and Lanaridi [38]. The influence of the composition of cations, anions, and organic solvents on the selectivity and completeness of extraction of the target metals, as well as possible extraction mechanisms depending on the forms of metal presence in solutions, are considered in detail.

Therefore, this chapter will cover just some recent examples of the following applications of ILs:

- as extractants;
- as solvents;
- synergetic mixtures of ILs;
- task-specific ionic liquids;
- as a phase-forming component or as an additive in aqueous biphasic systems.

3.1 ILs as extractants

A piperazine-based IL 1-(2-(dimethylamino) ethyl)-4-methyl-piperazin bis(trifluoromethylsulfonyl)imide ([C_6-Et-TMEDA-PIP][Tf_2N]$_2$) with two functional groups was synthesized to construct an extraction system for Au(III) and Pt(IV) recovery [39]. The anion exchange mechanism was verified through the combination of methods. The stripping procedure is carried out using $H_2C_2O_4$ and $CS(NH_2)_2$-HCl solutions for gold and platinum, respectively.

A commercially available quaternary ammonium salt Aliquate 336 has been used in a series of studies by Binnemans and coauthors dealing with different extraction approaches for precious metals separation [40–42]. Undiluted IL in the original chloride form [A336][Cl] or its substituted bromide form [A336][Br] enables separation of gold and palladium from base metals under nonequilibrium conditions in a milliflow set-up operating in the slug flow regime [41]. Another technique is a split-anion extraction with [A336][X] (X− = Br − and I−) developed for the separation of precious metals from aqueous chloride media. Ammonia solution, sodium thiosulfate, and thiourea were used for the selective stripping of Pd(II), Au(III), and Pt(IV), respectively, from loaded [A336][I] phase [42].

Deng et al. proposed a microemulsion extraction technique for palladium recovery from alkaline cyanide solutions using imidazolium ILs [43]. 1-butyl-3-undecyl imidazolium bromide ([BUIm]Br) was used as an extractant in a mixture of n-pentanol and n-heptane. Under optimal conditions, Pd(II) quantitatively transfers to the organic phase along with Fe(III) and Co(III) ions. The metals are separated through a two-step stripping procedure.

Quaternary ammonium chloride pseudo-protic ILs (PPILs) generated from the reaction of a primary, secondary, and tertiary amines and dissolved in toluene were used for gold(III) extraction from HCl media [44]. The metal could be conveniently precipitated as zero-valent gold nanoparticles after its stripping by sodium thiocyanate solution.

IL-based thermomorphic systems have been studied by Wang and coauthors for precious metals separation [45, 46]. UCST (upper critical solution temperature)-type

IL 1,4,7-trimethyltriazonane bis-(trifuloromethanesulfonyl) amide ([1,4,7-TMTA][Tf_2N]) saturated with water was used for homogenous liquid-liquid extraction of Au(III), Pd(II), and Pt(IV) at elevated temperature (40–65°C) [45]. Ethyl chloroacetate N,N,N′,N′-tetramethyl-ethylenediamine-based IL [EA-TMEDA][Tf_2N]$_2$ with temperature-responsive behavior revealed the high selectivity toward gold ions in the hydrochloric acid multicomponent solutions [46].

It should be noted that phosphonium-based ILs remain the most widely explored ILs in solvent extraction of platinum metals. Cyphos IL 101 has remarkable extraction ability to Pt(IV) and Pd(II) ions in a wide range of HCl concentrations and can be used for their recovery from various chloride-based leach solutions [47–49]. Other Cyphos ILs are full described in the above-afforded reviews [4, 5, 11, 36–38].

3.2 ILs as solvents

To separate PGMs from simulated high-level liquid waste (HLLW), a novel system containing N,N′-dimethyl-N,N′-di-(2-phenylethyl)-thiodiglycolamide (MPE-TDGA) as extractant and 1-butyl-3-methyl-imidazolium nonafluorobutansulfonate ([Bmim][NfO]) as a solvent was proposed [50]. The system allowed the rapid and selective extraction of Pd(II), as well as Ru(III) and Rh(III). These extractions were accelerated by increasing the temperature (50°C).

3.3 Synergetic mixtures of ILs

Under carefully selected experimental conditions, the distribution coefficients of some metals in systems containing a mixture of two extractants are much higher than the additive distribution coefficient of individual extractants. This phenomenon, called synergism, is due to, as it is supposed, different types of interactions between two extractants and/or the formation of metal-containing compounds of a special composition that differs from the composition of compounds in systems with one type extractant.

A synergistic effect in the extraction of precious metals is manifested when using the mixtures of hydrophilic and hydrophobic ILs.

Chen et al. developed extraction-electrodeposition method for platinum recovery from the multicomponent solutions using the mixture of ILs [C_{14}PIm][Br]/[C_8MIm][PF_6] (1-tetradecyl-3-propylimidazolium bromide/1-octyl-3-methylimidazolium hexafluorophosphate) [51]. Pt(IV) ions were selectively separated from metal solutions containing Rh^{3+}, Fe^{3+}, Ni^{2+}, Cu^{2+}, and Zn^{2+} metal ions, followed by direct electrodeposition as Pt(0) on the copper cathode.

Betaine-based IL [C_6Bet]Br was firstly applied to the separation of platinum metals [52]. [C_6Bet]Br showed remarkable extractability for Pt(IV) and Ir(IV) in the presence of 1-hexyl-3-methylimidazolium bis[(trifluoromethyl)sulfonyl]imide [C_6MIm][NTf_2] as the hydrophobic phase. Ir(IV) was separated from Pt(IV) by reduction process with the use of hydroxylamine hydrochloride ($NH_2OH{\cdot}HCl$).

The synergetic mixture of the imidazolium-based ILs [C_6MIm]Cl, [C_6MIm][NTf_2], and [C_6MIm][DDTC] (1-octyl-3-methylimidazolium chloride, bis[(trifluoromethyl)sulfonyl]imide and diethyldithiocarbamate, respectively) was offered for the extraction of Pt(IV), Pd(II), Ru(III), and Rh(III) [53]. Separation of these four PGMs can be realized by changing the composition of the synergetic mixture and concentrations of ILs.

3.4 Task-specific ionic liquids (TSILs)

These have functional group(s) involving donor atom(s) to selectively extract the targeted metal ions through metal complex formation.

A pyridinium-based TSIL with a monothioether group, 3-thiapentylpyridinium bis(trifluoromethylsulfonyl)imide [3-TPPy][NTf_2], was prepared for the extraction of typical class b metal ions, including precious metal ions in high selectivity [54].

A novel extraction-photocatalysis method was developed to recover Pd from HCl leaching solutions using [C_5MIm][DDTC] (1-pentyl-3-methylimidazolium diethyldithiocarbamate) as extractant [55]. The extraction was considered due to Pd-S coordination. Direct stripping of the metal can be accomplished through photocatalytic reduction process.

The TSILs bearing one or two tetrahydropyran-2H-yl(THP)-protected thiols were designed as palladium extractants from an aqueous phase to [C_4MIm][NTf_2] [56]. Such a kind of synergetic mixture allows to selectively bind Pd(II) ions from 4 M HCl in the presence of Pt(IV).

2-Mercaptobenzothiazole-functionalized IL ([C_6mim][2MBT]) in combination with the nonionic surfactant (TX-114) was used for palladium separation *via* cloud point extraction [57]. The obtained results showed a strong coordination of palladium with anion of the IL.

An urea-based imidazolium IL, 1-butyl-3-{3-(3-methyl-2H-imidazol-1-yl)propyl} urea bis(trifluoromethylsulfonyl)imide ([C_4UC_3mim][Tf_2N]), was synthesized and studied for Pt(IV), Pd(II) and Rh(III) separation [58]. Based on experimental data, the anion exchange mechanism was supposed for the extraction of Pt(IV) at pH = 1.13 and the inner-sphere coordination for Pd(II) extraction at pH = 5.45.

3.5 ILs as a phase-forming component or as an additive in aqueous biphasic systems

An IL-based ABSs can overcome the limitations of the use of ILs associated with their high viscosity and hydrophobicity [59]. A novel type of acidic aqueous biphasic systems (AcABS) based on an inorganic acid and an IL has been recently gaining interest for the extraction of metal ions as a promising alternative to conventional extraction systems [60]. The AcABS composed from hydrochloric acid with high concentrations and phosphonium-based IL ([P_{44414}][Cl]) is characterized by thermotropic behavior and can be successfully applied to the separation of some strategic metals, including Pt(IV) [60].

3.6 Other application of ILs

Solid-phase extraction and membrane separation have such advantages as lower consumption of IL and its improved stability. Both technologies are considered as promising alternatives to traditional solvent extraction and deserve a separate chapter. Here are just a few references on the use of ILs as functionalizing modifiers of various solid carriers: resins, membranes, and nanoparticles [61–66].

One of the recent areas of research should also be noted. It is related to the leaching of platinum-containing materials using ILs. This approach is intended as an alternative to the refining processes, where strong acids with oxidants (aqua regia, concentrated HCl in the presence of chlorine gas or H_2O_2) are usually used to dissolve chemically inert platinum metals. Thus, a nonaqueous direct leaching process based on trihexyl(tetradecyl)phosphonium chloride ($P_{66614}Cl$) in combination with methanesulfonic acid and trichloroisocyanuric acid was proposed to solubilize metallic platinum [67].

4. Deep eutectic solvents

Since the first publication about 20 years ago by Abbott [68], deep eutectic solvents (DESs) have evolved from a subclass of ionic liquids to a wide range of liquids that have applications in many fields, including analytical chemistry and separation techniques. The original description of DES as a mixture of a salt of a quaternary ammonium base (a hydrogen bond acceptor, HBA) and a hydrogen bond donor (HBD) has been recently extended to a combination of Lewis and Bronsted acids and bases with much lower melting temperatures of the resulting mixture compared to the that of the original components [69].

DESs have a number of advantages: easy and convenient preparation, non-toxicity, non-volatility, non-combustibility, and biodegradability, which allow these compounds to be classified as environmentally benign [19, 70]. As in the case of ILs, due to the variability of constituent components, DESs are referred to designer solvents. In the emerging era of the transfer to active use of artificial intelligence, the selection of a certain composition of DESs using neural networks seems to be an extremely interesting and promising area of research [71, 72].

Publications on precious metal extraction by DESs are summarized in **Table 2**, focusing on the selected DES composition and extraction conditions. One of the pioneering works on DES application in the separation of precious metals appeared in 2019 and was devoted to the extraction of gold. Geng et al. studied the DESs based on quaternary ammonium salts for this purpose [73]. The hydrophobic DESs with $[N_{3333}]Br$, $[N_{4444}]Br$ and $[N_{8881}]Br$ as HBAs and N-hexanoic acid as HBD (1:1 molar ratio) were screened out for gold recovery from hydrochloric acid solutions with various acidity and salinity. Based on combination of UV–Vis and FT-IR spectroscopy data, the anion exchange mechanism was proposed: anions $[C_5H_{13}\text{-}COOH\cdots Br]^-$ are exchanged with $AuCl_4^-$ during the extraction process. The outcomes under different conditions showed that $N_{8881}Br$ has the best gold extraction ability and tolerance to higher salinity. $NaBH_4$ solution (0.1 mg/L) was used as a stripping agent. The gold extraction ability of DESs was preserved during 5 cycles of extraction.

Choline chloride: phenol mixture in 1:2 molar ratio was introduced by Yilmaz et al. for gold extraction [74]. The goal of the work was to develop a sensitive method for determination of gold traces in plating bath solutions. The combined approach was used to achieve this objective: liquid phase microextraction by DES in the presence of a complexing agent and detection by flame atomic absorption spectrometry with a slotted quartz tube (SQT-FAAS). Additionally, the conventional parameters of the extraction efficiency were studied as a function of the mixing mode. Out of hand mixing, mechanical shaking, ultrasonic bath, and vortexing, the latter one was selected as the most efficient. For better phase separation, tetrahydrofuran was used for the emulsification of DES.

A similar approach and a DES of the same type were used by Panhwar et al. for palladium extraction coupled with FAAS determination in environmental samples [75]. Choline chloride:phenol mixture in 1:4 molar ratio, 2-hydroxy-3-methoxy benzaldehyde thiosemicarbazone as a complexing agent, and tetrahydrofuran as an emulsifier were dispersed in sample solution by 8 repeated cycles of quick uptaking and discharging with the use of a syringe. Selectivity of Pd(II) extraction was evaluated in the presence of Co(II), Cu(II), Cd(II), Ni(II), Mn(II), Zn(II), Pb(II), and Cr(III) ions.

Metal	DES composition	Extraction conditions	Stripping/ determination conditions	Reference
Au(III)	HBA: $N_{8881}Br$, HBD: N-hexanoic acid	pH = 2, 1 mL of 1 mM DES, t = 30 min, Sample V = 1 mL, C_0 = 0.2–0.6 mM (39–118 $\mu g{\cdot}mL^{-1}$)	0.1 $g{\cdot}mL^{-1}$ $NaBF_4$, V = 1 mL	[73]
Au(III)	HBA: choline chloride, HBD: phenol	pH = 6, 1 mL of 0.05% sodium diethyl dithiocarbamate, 500 μL of DES, t = 15 s, V = 10 mL, C_0 = 500 $\mu g{\cdot}L^{-1}$; 0.5 mL of THF as emulsifier	FAAS determination in 100 μL of DES, LOD = 5.1 $\mu g{\cdot}L^{-1}$	[74]
Pd(II)	HBA: choline chloride, HBD: phenol	pH = 5, 400 μL of 0.1% HMBATSC*, 500 μL of DES, V = 35 mL, C_0 = 10 $\mu g{\cdot}L^{-1}$; 0.8 mL of THF as emulsifier	FAAS determination in 100 μL of DES, LOD = 4.0 $\mu g{\cdot}L^{-1}$	[75]
Pd(II)	HBA: hydroxyl ammonium chloride, HBD: phenol, and disodium 4,5-dihydroxy-1,3-benzenedisulfonate; $FeCl_3$	pH = 6, 100 μL of DES, t = 1 min, V = 10–35 mL	FAAS determination in DES diluted by 0.3 mL of conc. HNO_3, LOD = 1.18 $\mu g{\cdot}L^{-1}$	[76]
Pd(II)	HBA: DL-menthol, HBD: phenyl salicylate (salol)	pH = 4, $1.5{\cdot}10^{-5}$ M PAN, 60 μL of DES, T = 65°C, t = 1 min, V = 10 mL	ETAAS determination in 20 μL of DES, LOD = 0.03 $\mu g{\cdot}L^{-1}$	[77]
Pd(II)	HBA: $N_{8881}Cl$, HBD: ethylene glycol, glycerol, 1-hexanol, C_2H_5COOH, $C_5H_{11}OOH$	0.1–0.3 M HCl, m = 0.05 g, V = 5 mL, t = 20 min, C_0 = 60–350 $\mu g{\cdot}mL^{-1}$	80 wt% $N_2H_4{\cdot}H_2O$, V = 0.5 mL	[78]
Pd(II), Pt(IV)	HBA: $N_{8888}Br$, HBD: octanoic acid	0.01–6 M HCl, m = 0.05–1.5 g, V = 1.5–10 mL, t = 10 min, C_0 = 5–10 $\mu g{\cdot}mL^{-1}$	0.25 M NH_4OH (Pd), 2 M HNO_3 (Pt)	[79]
Pt(IV)	HBA: TOPO, HBD: 1-butanol	1–4 M HCl, m = 0.042 g, V = 3 mL, t = 60 min, C_0 = 682–1092 $\mu g{\cdot}mL^{-1}$	0.5 M NaOH, V = 3 mL	[80]
Pt(IV)	HBA: TOPO, HBD: thymol	2 M HCl, O/A = 1 (vol/vol, total volume 2.0 mL), t = 1 min, C_0 = 390 $\mu g{\cdot}mL^{-1}$	0.1 M thiourea in 0.5 M HCl, V = 1 mL	[81]

**HMBATSC – 2-hydroxy-3-methoxy benzaldehyde thiosemicarbazone.*

Table 2.
The examples of DES application for the extraction of precious metals.

Four-component $FeCl_3$-based DES was applied by ALOthman et al. for dispersive microextraction of palladium followed by FAAS determination [76]. To prepare the DES disodium 4,5-dihydroxy-1,3-benzenedisulfonate, hydroxylammonium chloride, iron(III) chloride and phenol were mixed in the ratio of 1:1:2:1. In addition to typical hydrogen-bonding interactions, the presence of $FeCl_3$ in the formulation leads to coordination interactions with oxygen atoms of donor ligands ensuring the formation of a liquid. The developed procedure does not require a complexing agent; palladium recovery is quantitative and tolerant to the presence of common matrix ions.

Another kind of DES was proposed for palladium extraction by Abdi. et al. [77]. DL-menthol was mixed with phenyl salicylate (1:1), and the prepared DES was dispersed in the sample solution together with 1-(2-pyridylazo)-2-naphtol complexing agent. The extraction procedure was carried out at an elevated temperature (65°C), when the homogeneous solution was formed. The extremely low limit of detection could be achieved using the ETAAS determination method.

Abovementioned long-chain quaternary ammonium salt was investigated by Tang. et al. for the extraction of palladium from acid solutions [78]. DES was constructed by combining $N_{8881}Cl$ with saturated fatty acids or fatty alcohols in molar ratio 1:1 and 1:2. An anion exchange mechanism of extraction was confirmed by FTIR, UV-Vis, and 1H NMR analysis. Authors proposed stripping of palladium from the DES phase via the hydrazine hydrate reduction method; the regenerated DES was used repeatedly for Pd(II) extraction for 5 cycles.

Extraction of the whole group of platinum group metals from acid solutions with high amount of chloride-ions and matrix components was studied using DESs based on tetraoctylammonium bromide ($N_{8888}Br$) and carboxylic acids [79]. The scheme of separation of Pd(II) and Pt(IV) as well as rare platinum metals was proposed for processing technological solutions to obtain individual fractions of precious metals with purity of >99.9%.

Trioctylphosphine oxide (TOPO) as HBA was mixed with various HBD reagents by Liu et al. to construct the extraction systems for Pt(IV) recovery from secondary resources [80]. TOPO-1-butanol, TOPO-L-menthol, and TOPO-1-hexanol were recommended for Pt(IV) extraction under conditions of high acidity and salinity. The ion-association mechanism of extraction was elucidated: each of the two protonated P=O groups in the TOPO molecule combined with $PtCl_6^{2-}$ during the process of extraction. The TOPO-1-butanol possessed the best Pt(IV) extraction ability, selectivity, and cycling extraction (using NaOH as a stripping agent).

One of the most critical and comprehensive study was carried out by Vargas, Schaeffer, and their colleagues [81, 82]. The work provides the features of Pt(IV) and Pd(II) extraction using TOPO-based DES in comparison with an equivalent extractant system in organic diluent. The conclusions of this work are quite important and deserve to be summarized here [81, 82]:

- on a molecular level, in the case of TOPO and diluent organic system, less molecules of extractant are required to form metal complexes in comparing with TOPO-decanoic acid DES;

- there is the competitive intermolecular hydrogen bonding between the DES components and TOPO···H^+ adduct formation, which precedes the metal extraction; the higher HBD concentration in the DES, the higher HCl concentration, at which metal extraction occurs;

- due to extensively developed hydrogen bonded network, the DES extraction system does not form the third phase even at 8 M HCl in contrast to extractant in diluents;

- Pt(IV)/Pd(II) separation factor could be tuned by HBD selection, for example, at lower HCl concentrations in the TOPO-thymol system Pd(II) extraction is inhibited, while Pt(IV) extraction remains unaffected.

The reviewed publications demonstrate mainly the possibility of using DESs in analytical methods for the determination of three metals: gold, platinum, and palladium. The components of the studied DESs used as HBAs, namely the quaternary ammonium salts and TOPO, have been well studied in the traditional extraction of platinum metals. If the use of DESs is to be positioned as a transition to more environmentally friendly technologies, then DESs based on nontoxic compounds like choline chloride and menthol should be given a preference. However, in the case of such "green" components of the DESs, a complex-forming agent needs to be added to the extraction system [74, 75]. In order not to complicate the system and the subsequent detection stage, the design of DESs using components, which are able to complex-forming or specific interaction with platinum metals for their selective recovery, is a subject of a scientific interest. The recent work, authored by Liu, has been devoted to Au(III), Pd(II), and Pt(IV) extraction using the eutectic mixture of natural components: lidocaine and thymol [83]. Thus, the search for safe and efficient eutectic extraction systems for precious metals can be expected to continue.

5. Conclusion

The imperative need to improve the existing and create new schemes for separation of PGMs and gold from complex matrices concerns both technological processes and the analytical control of the metals contents at all production stages. In many cases, selective separation of precious metals is necessary, and this is the most difficult stage of processing platinum-containing materials owing to their multicomponent composition, specific chemistry of PGMs, and their extremely low concentrations. Conventional leaching and extraction methods for the separation of precious metals involve the use of aggressive reagents and toxic solvents and are characterized by high energy consumption. The presented review confirms the relevance of fundamental and applied research in the use of alternative solvents for separation of these critical elements. At the moment, the processing of technological wastes, spent catalysts, and other secondary resources is of priority importance because of their large quantities and depletion of natural resources. In this regard, the problem of precious metals production, on the one hand, becomes a global challenge and, on the other hand, requires innovative technological solutions for countries that have primary raw materials and operate outdated approaches and materials.

Applications of ABSs, ILs, and DESs in precious metals' recycling benefit from their excellent solubility properties, extraction and electrochemical activities, and eco-friendly portfolio. Further studies in this field are likely to focus on the search for green selective components with high affinity to precious metals, especially to rare platinum metals (rhodium, iridium, and ruthenium), including the computational research and design of extraction systems, combining extraction and leaching procedures with subsequent determination, or recovery of precious metals from leachate using simple and sustainable approaches.

Acknowledgements

This work was supported by the Ministry of Science and High Education of the Russian Federation [GEOKHI RAS].

Author details

Olga Mokhodoeva
Vernadsky Institute of Geochemistry and Analytical Chemistry, Russian Academy of Sciences, Moscow, Russia

*Address all correspondence to: olga.mokhodoeva@mail.ru

References

[1] The Johnson Matthey PGM Market Report. 2022. Available from: https://matthey.com/pgm-market-report-2022 [Accessed: May 16, 2022]

[2] Tang H, Peng Z, Tian R, Ye L, Zhang J, Rao M, et al. Platinum-group metals: Demand, supply, applications and their recycling from spent automotive catalysts. Journal of Environmental Chemical Engineering. 2023;**11**:110237. DOI: 10.1016/j.jece.2023.110237

[3] Hughes AE, Haque N, Northey SA, Giddey S. Platinum group metals: A review of resources, production and usage with a focus on catalysts. Resources. 2021;**10**:93. DOI: 10.3390/resources10090093

[4] Xia J, Ghahreman A. Platinum group metals recycling from spent automotive catalysts: Metallurgical extraction and recovery technologies. Separation and Purification Technology. 2023;**311**:123357. DOI: 10.1016/j.seppur.2023.123357

[5] Zheng H, Ding Y, Wen Q, Liu B, Zhang S. Separation and purification of platinum group metals from aqueous solution: Recent developments and industrial applications. Resources, Conservation and Recycling. 2021;**167**:105417. DOI: 10.1016/j.resconrec.2021.105417

[6] Sinisalo P, Lundström M. Refining approaches in the platinum group metal processing value chain—A review. Metals. 2018;**8**:203. DOI: 10.3390/met8040203

[7] Saguru C, Ndlovu S, Moropeng D. A review of recent studies into hydrometallurgical methods for recovering PGMs from used catalytic converters. Hydrometallurgy. 2018;**182**:44-56. DOI: 10.1016/j.hydromet.2018.10.012

[8] Komendova R. Recent advances in the preconcentration and determination of platinum group metals in environmental and biological samples. Trends in Analytical Chemistry. 2020;**122**:115708. DOI: 10.1016/j.trac.2019.115708

[9] Crundwell FK, Moats MS, Ramachandran V, Robinson TG, Davenport WG. Chapter 37—Refining of the platinum-group metals. In: Crundwell FK, Moats MS, Ramachandran V, Robinson TG, Davenport WG, editors. Extractive Metallurgy of Nickel, Cobalt and Platinum Group Metals. Oxford: Elsevier; 2011. pp. 489-534. DOI: 10.1016/B978-0-08-096809-4.10037-1

[10] Liu G, Wu Y, Tang A, Pan D, Li B. Recovery of scattered and precious metals from copper anode slime by hydrometallurgy: A review. Hydrometallurgy. 2020;**197**:105460. DOI: 10.1016/j.hydromet.2020.105460

[11] Pianowska K, Kluczka J, Benke G, Goc K, Malarz J, Ochmański M, et al. Solvent extraction as a method of recovery and separation of platinum group metals. Materials. 2023;**16**:4681. DOI: 10.3390/ma16134681

[12] Mokhodoeva OB, Myasoedova GV, Kubrakova IV. Sorption preconcentration in combined methods for the determination of noble metals. Journal of Analytical Chemistry. 2007;**62**(7):607-622. DOI: 10.1134/S1061934807070027

[13] Rao CRK, Trivedi DC. Chemical and electrochemical depositions of platinum

group metals and their applications. Coordination Chemistry Reviews. 2005;**249**(5-6):613-631. DOI: 10.1016/j.ccr.2004.08.015

[14] Izatt RM, Izatt SR, Izatt NE, Krakowiak KE, Bruening RL, Navarro L. Industrial applications of molecular recognition technology to separations of platinum group metals and selective removal of metal impurities from process streams. Green Chemistry. 2015;**17**:2236-2245. DOI: 10.1039/C4GC02188F

[15] Massoud M, Vega G, Subburaj A, Partheepan J. Review on recycling energy resources and sustainability. Helyon. 2023;**9**:e15107. DOI: 10.1016/j.heliyon.2023.e15107

[16] Zupanc A, Install J, Jereb M, Repo T. Sustainable and selective modern methods of noble metal recycling. Angewandte Chemie, International Edition. 2023;**62**:e202214453. DOI: 10.1002/anie.202214453

[17] Ilyas S, Kim H, Srivastava RR, editors. Sustainable Urban Mining of Precious Metals. 1st ed. London, UK: CRC Press; 2021. p. 262. DOI: 10.1201/9781003055068

[18] Karmakar R, Sen K. Aqueous biphasic extraction of metal ions: An alternative technology for metal regeneration. Journal of Molecular Liquids. 2019;**273**:231-247. DOI: 10.1016/j.molliq.2018.10.036

[19] Pletnev IV, Smirnova SV, Sharov AV, Zolotov YA. Next-generation extraction solvents: From ionic liquids to aqueous biphasic systems to low-melting combined solvents. Russian Chemical Reviews. 2021;**90**(9):1109-1141. DOI: 10.1070/RCR5007

[20] Albertsson P-A. Partition of Cell Particles and Macromolecules. 3rd ed. New York: Wiley; 1986. p. 346

[21] Diamond AD, Hsu JT. Aqueous two-phase systems for biomolecule separation. In: Tsao GT, editor. Bioseparation. Advances in Biochemical Engineering/Biotechnology. Vol. 47. Berlin, Heidelberg: Springer; 1992. DOI: 10.1007/BFb0046198

[22] Saha N, Sarkar B, Sen K. Aqueous biphasic systems: A robust platform for green extraction of biomolecules. Journal of Molecular Liquids. 2022;**363**:119882. DOI: 10.1016/j.molliq.2022.119882

[23] Zvarova TI, Shkinev VM, Vorob'eva GA, Spivakov BYa, Zolotov YuA. Liquid-liquid extraction in the absence of usual organic solvents: Application of two-phase aqueous systems based on a water-soluble polymer. Microchimica Acta. 1984;**84**:449-458. DOI: 10.1007/BF01197159

[24] Bulgariu L, Bulgariu D. Extraction of gold(III) from chloride media in aqueous polyethylene glycol-based two-phase system. Separation and Purification Technology. 2011;**80**:620-625. DOI: 10.1016/j.seppur.2011.06.018

[25] Simonova TN, Dubrovina VA. Palladium(II) chloride complexes extraction in aqueous segregable systems with following spectrophotometric determination. Proceedings of Voronezh State University. Series: Chemistry. Biology. Pharmacy. 2018;**4**:42-47. Available from: www.vestnik.vsu.ru/pdf/chembio/2018/04/2018-04-05.pdf

[26] Milevskiy N, Boryagina I, Karpukhina E, Kuznetsov V. Effect of sodium chloride and pH on the composition of the equilibrium phases and the partition of palladium(II) in the aqueous two-phase system PEG1500–Na2SO4–Water. Journal of Chemical & Engineering Data. 2021;**66**:1021-1031. DOI: 10.1021/acs.jced.0c00832

[27] Hespanhol MC, Fontoura BM, Quintão JC, da Silva LHM. Extraction and purification of gold from raw acidic electronic leachate using aqueous biphasic system. Journal of the Taiwan Institute of Chemical Engineers. 2020;**15**:218-222. DOI: 10.1016/j.jtice.2020.10.027

[28] Zheng Y, Tong Y, Wang S, Zhang H, Yang Y. Mechanism of gold(III) extraction using a novel ionic liquid-based aqueous two phase system without additional extractants. Separation and Purification Technology. 2015;**154**:123-127. DOI: 10.1016/j.seppur.2015.09.014

[29] Tang N, Wang Y, Liu M, Liu L, Yin C, Yang X, et al. Ionic liquid as adjuvant in an aqueous biphasic system composed of polyethylene glycol for green separation of Pd(II) from hydrochloric solution. Separation and Purification Technology. 2020;**246**:116898. DOI: 10.1016/j.seppur.2020.116898

[30] Mokhodoeva O, Rudik I, Shkinev V, Maryutina T. Countercurrent chromatography approach to palladium and platinum separation using aqueous biphasic system. Journal of Chromatography. A. 2021;**1657**:462581. DOI: 10.1016/j.chroma.2021.462581

[31] Simonova TN, Nekrasova EA. Extraction by separation of multi-charge chloride complexes of iridium(IV) and rhodium(III) in segregated water systems. Sorbtsionnye i Khromatograficheskie Protsessy. 2020;**20**(2):299-304. DOI: 10.17308/sorpchrom.2020.20/2785

[32] Kaur G, Kumar H, Singla M. Diverse applications of ionic liquids: A comprehensive review. Journal of Molecular Liquids. 2022;**351**:118556. DOI: 10.1016/j.molliq.2022.118556

[33] Inman G, Nlebedim I, Prodius D. Application of ionic liquids for the recycling and recovery of technologically critical and valuable metals. Energies. 2022;**15**:628. DOI: 10.3390/en15020628

[34] Chen Y, Mu T. Revisiting greenness of ionic liquids and deep eutectic solvents. Green Chemical Engineering. 2021;**2**:174-186. DOI: 10.1016/j.gce.2021.01.004

[35] Bystrzanowska M, Pena-Pereira F, Marcinkowski Ł, Tobiszewski M. How green are ionic liquids?—A multicriteria decision analysis approach. Ecotoxicology and Environmental Safety. 2019;**174**:455-458. DOI: 10.1016/j.ecoenv.2019.03.014

[36] Lee J, Kurniawan K, Kim S, Nguyen VT, Pandey BD. Ionic liquids-assisted solvent extraction of precious metals from chloride solutions. Separation and Purification Reviews. 2022;**52**(3):242-261. DOI: 10.1080/15422119.2022.2091458

[37] Firmansyah ML, Yoshida W, Hanada T, Goto M. Application of ionic liquids in solvent extraction of platinum group metals. Solvent Extraction Research and Development, Japan. 2020;**27**(1):1-24. DOI: 10.15261/SERDJ.27.1

[38] Lanaridi O, Schnürch M, Limbeck A, Schröder K. Liquid- and solid-based separations employing ionic liquids for the recovery of platinum group metals typically encountered in catalytic converters: A review. ChemSusChem. 2022;**15**(6):e202102262. DOI: 10.1002/CSSC.202102262

[39] Wang Y, Liu X, Yang A, Lv P, Zhang L, Li Y, et al. Extraction and separation of Au(III) and Pt(IV) from HCl media using novel piperazine-based ionic liquid as an ionic exchanger. Journal of Molecular Liquids. 2022;**353**:118846. DOI: 10.1016/j.molliq.2022.118846

[40] Vereycken W, Riaño S, Van Gerven T, Binnemans K. Extraction behavior and separation of precious and base metals from chloride, bromide, and iodide media using undiluted halide ionic liquids. ACS Sustainable Chemistry & Engineering. 2020;**8**(22):8223-8234. DOI: 10.1021/acssuschemeng.0c01181

[41] Vereycken W, Van Stee J, Riaño S, Van Gerven T, Binnemans K. Non-equilibrium solvent extraction in milliflow reactors: Precious and base metal separations with undiluted ionic liquids. Separation and Purification Technology. 2021;**265**:118490. DOI: 10.1016/j.seppur.2021.118490

[42] Nguyen VT, Riaño S, Binnemans K. Separation of precious metals by split-anion extraction using water-saturated ionic liquids. Green Chemistry. 2020;**22**:8375-8388. DOI: 10.1039/D0GC02356F

[43] Deng H, Liu C, Xu X, Wu Y, Chen M, Huang Z. Separation of palladium from alkaline cyanide solutions through microemulsion extraction using imidazolium ionic liquids. International Journal of Molecular Sciences. 2023;**24**:10709. DOI: 10.3390/ijms241310709

[44] Alguacil FJ, Robla JI. On the use of pseudo-protic ionic liquids to extract gold(III) from HCl solutions. International Journal of Molecular Sciences. 2023;**24**:6305. DOI: 10.3390/ijms24076305

[45] Wang Y, Chen S, Liu R, Zhang L, Xue W, Yang Y. Toward green and efficient recycling of Au(III), Pd(II) and Pt(IV) from acidic medium using UCST-type ionic liquid. Separation and Purification Technology. 2022;**298**:121620. DOI: 10.1016/j.seppur.2022.121620

[46] Wang Y, Liu R, Liu X, Lv P, Zhang L, Yang Y. Synthesis of temperature-responsive ionic liquids and study on the behavior and mechanism of Au(III) extraction from hydrochloric acid solutions. Hydrometallurgy. 2022;**211**:105859. DOI: 10.1016/j.hydromet.2022.105859

[47] Ilyas S, Kim H, Srivastava RR. Separation of platinum group metals from model chloride solution using phosphonium-based ionic liquid. Separation and Purification Technology. 2021;**278**:119577. DOI: 10.1016/j.seppur.2021.119577

[48] Wiecka Z, Rzelewska-Piekut M, Regel-Rosocka M. Recovery of platinum group metals from spent automotive converters by leaching with organic and inorganic acids and extraction with quaternary phosphonium salts. Separation and Purification Technology. 2022;**280**:119933. DOI: 10.1016/j.seppur.2021.119933

[49] Paiva AP, Piedras FV, Rodrigues PG, Nogueira CA. Hydrometallurgical recovery of platinum-group metals from spent auto-catalysts—Focus on leaching and solvent extraction. Separation and Purification Technology. 2022;**286**:120474. DOI: 10.1016/j.seppur.2022.120474

[50] Oosugi H, Ito T, Takahashi T, Wu H, Kim S-Y. Extraction behaviors of platinum group metals from an aqueous HNO_3 solution using ionic liquids containing a novel thiodiglycolamide-type extractant. Journal of Radioanalytical and Nuclear Chemistry. 2022;**331**:4577-4585. DOI: 10.1007/s10967-022-08564-y

[51] Chen M, Li S, Jin C, Shao M, Huang Z. Selective recovery of platinum by combining a novel reusable ionic liquid with electrodeposition. Separation and Purification Technology. 2021;**259**:118204. DOI: 10.1016/j.seppur.2020.118204

[52] Yan Y, Wang Q, Xiang Z, Zheng Y, Wang S, Yang Y. Behavior and mechanism investigation of separating Pt and Ir by liquid-liquid extraction using a mixed [C_6bet]Br/[C_6mim][NTf_2] system. New Journal of Chemistry. 2017;**41**(17):8985-8992. DOI: 10.1039/C7NJ02361H

[53] Yan Y, Wang Q, Xiang Z, Yang Y. Separation of Pt(IV), Pd(II), Ru(III), and Rh(III) from chloride medium using liquid–liquid extraction with mixed imidazolium-based ionic liquids. Separation Science and Technology. 2018;**53**(13):2064-2073. DOI: 10.1080/01496395.2018.1440304

[54] Chayama K, Sano Y, Iwatsuki S. Pyridinium-based task-specific ionic liquid with a monothioether group for selective extraction of class b metal ions. Analytical Sciences. 2015;**31**:1115-1117. DOI: 10.2116/analsci.31.1115

[55] Chen M, Li S, Luo G, Fan M, Chen J, Huang Z, et al. Selective recovery of palladium and rhodium by combined extraction and photocatalytic reduction. Separation and Purification Technology. 2021;**274**:119006. DOI: 10.1016/j.seppur.2021.119006

[56] Funaki K, Ma S, Kawamura S, Miyazaki A, Sugie A, Mori A, et al. Metal-selective deprotection-mediated palladium(II) extraction by ionic liquids with tetrahydropyran-2H-yl-protected thiol moieties. Chemistry Letters. 2017;**46**:434-437. DOI: 10.1246/cl.161153

[57] Wu Q, Zhao C, Wang Q, Liu J. Successive preconcentration and mechanistic investigation of Au(III), Pd(II), Pt(IV) and Rh(III) via cloud point extraction using a functionalised ionic liquid. New Journal of Chemistry. 2022;**46**:7695. DOI: 10.1039/d2nj00191h

[58] Ueda Y, Eguchi A, Tokunaga K, Kikuchi K, Sugita T, Okamura H, et al. Urea-introduced ionic liquid for the effective extraction of Pt(IV) and Pd(II) ions. Industrial and Engineering Chemistry Research. 2022;**61**(19):6640-6649. DOI: 10.1021/acs.iecr.2c00304

[59] Ola PD, Matsumoto M. Metal extraction with ionic liquids-based aqueous two-phase system. Chapter 8. In: Rahman MM, editor. Recent Advances in Ionic Liquids. London, UK: InTech; 2018. DOI: 10.5772/intechopen.77286

[60] Gras M, Papaiconomou N, Schaeffer N, Chainet E, Tedjar F, Coutinho JAP, et al. Ionic-liquid-based acidic aqueous biphasic systems for simultaneous leaching and extraction of metallic ions. Angewandte Chemie, International Edition. 2018;**57**:1563-1566. DOI: 10.1002/anie.201711068

[61] Navarro R, Lira MA, Saucedo I, Alatorre A, Guibal E. Amberlite XAD-1180 impregnation with Cyphos IL101 for the selective recovery of precious metals from HCl solutions. Gold Bulletin. 2017;**50**(1):7-23. DOI: 10.1007/S13404-016-0190-8

[62] Cocheci L, Lupa L, Tolea NS, Lazău R, Pode R. IL-functionalized Mg_3Al-LDH as new efficient adsorbent for Pd recovery from aqueous solutions. International Journal of Molecular Sciences. 2022;**23**(16):9107. DOI: 10.3390/IJMS23169107

[63] Zhou S, Song N, Lv X, Jia Q. Magnetic dual task-specific polymeric ionic liquid nanoparticles for preconcentration and determination of gold, palladium and platinum prior to their quantitation by graphite furnace AAS. Microchimica Acta. 2017;**184**:3497-3504. DOI: 10.1007/s00604-017-2354-5

[64] Mokhodoeva O, Shkinev V, Maksimova V, Dzhenloda R, Spivakov B. Recovery of platinum group metals using

magnetic nanoparticles modified with ionic liquids. Separation and Purification Technology. 2020;**248**:117049. DOI: 10.1016/j.seppur.2020.117049

[65] Dzhenloda RK, Mokhodoeva OB, Danilova TV, Maksimova VV, Shkinev VM. Ultrasound suspension columns for solid-phase extraction of platinum and palladium. In: Advances in Geochemistry and Analytical Chemistry—Special Publication to 75th Anniversary of Vernadsky Institute of Geochemistry and Analytical Chemistry of the Russian Academy of Sciences. Cham, Switzerland: Springer International Publishing; 2023. pp. 535-540. DOI: 10.1007/978-3-031-09883-3_32

[66] Fajar ATN, Hanada T, Goto M. Recovery of platinum group metals from a spent automotive catalyst using polymer inclusion membranes containing an ionic liquid carrier. Journal of Membrane Science. 2021;**629**:119296. DOI: 10.1016/j.memsci.2021.119296

[67] Kamisono M, Hanada T, Goto M. Green leaching of metallic platinum using an ionic liquid with synergistic organic acid–chlorinating agent additives. Journal of Chemical Engineering of Japan. 2023;**56**(1):2228363. DOI: 10.1080/00219592.2023.2228363

[68] Abbot AP, Capper GC, Davies DL, Rasheed RK, Tambyrajah V. Novel solvent properties of choline chloride/urea mixtures. Chemical Communications. 2003;(1):70-71. DOI: 10.1039/B210714G

[69] Abbot AP, Edler KJ, Page AJ. Deep eutectic solvents—The vital link between ionic liquids and ionic solutions. The Journal of Chemical Physics. 2021;**155**:150401. DOI: 10.1063/5.0072268

[70] Yuan Z, Liu H, Yong WF, She Q, Esteban J. Status and advances of deep eutectic solvents for metal separation and recovery. Green Chemistry. 2022;**24**:1895-1929. DOI: 10.1039/D1GC03851F

[71] Tolmachev D, Lukasheva N, Ramazanov R, Nazarychev V, Borzdun N, Volgin I, et al. Computer simulations of deep eutectic solvents: Challenges, solutions, and perspectives. International Journal of Molecular Sciences. 2022;**23**: 645. DOI: 10.3390/ijms23020645

[72] Abranches DO, Coutinho JAP. Type V deep eutectic solvents: Design and applications. Current Opinion in Green and Sustainable Chemistry. 2022;**35**:1006112. DOI: 10.1016/j.cogsc.2022.100612

[73] Geng Y, Xiang Z, Lv C, Wang N, Wang Y, Yang Y. Recovery of gold from hydrochloric medium by deep eutectic solvents based on quaternary ammonium salts. Hydrometallurgy. 2019;**188**:264-271. DOI: 10.1016/j.hydromet.2019.06.013

[74] Yilmaz Ö, Durak BY, Tekin Z, Koçoğlu ES, Bakirdere S. Accurate and precise determination of gold in plating bath solution: Deep eutectic solvent based liquid phase microextraction—Slotted quartz tube—Flame atomic absorption spectrometry. Analytical Letters. 2020;**53**:165-173. DOI: 10.1080/00032719.2019.1641718

[75] Panhwar AH, Tuzen M, Kazi TG. Use of deep eutectic solvent-based air-assisted emulsification liquid-liquid microextraction of palladium and determination by flame atomic absorption spectrometry in water and environmental samples. Atomic Spectroscopy. 2019;**40**:227-232. DOI: 10.46770/AS.2019.06.005

[76] ALOthman ZA, Habila MA, Yilmaz E, Alabdullkarem EA, Soylak M. A novel deep eutectic

solvent microextraction procedure for enrichment, separation and atomic absorption spectrometric determination of palladium at ultra-trace levels in environmental samples. Measurement. 2020;**153**:107394. DOI: 10.1016/j.measurement.2019.107394

[77] Abdi K, Ezoddin M, Pirooznia N. Temperature-controlled liquid–liquid microextraction using a biocompatible hydrophobic deep eutectic solvent for microextraction of palladium from catalytic converter and road dust samples prior to ETAAS determination. Microchemical Journal. 2020;**157**:104999. DOI: 10.1016/j.microc.2020.104999

[78] Tang N, Liu L, Yin C, Zhu G, Huang Q, Dong J, et al. Environmentally benign hydrophobic deep eutectic solvents for palladium(II) extraction from hydrochloric acid solution. Journal of the Taiwan Institute of Chemical Engineers. 2021;**121**:92-100. DOI: 10.1016/j.jtice.2021.04.010

[79] Mokhodoeva O, Maksimova V, Shishov A, Shkinev V. Separation of platinum group metals using deep eutectic solvents based on quaternary ammonium salts. Separation and Purification Technology. 2023;**305**:122427. DOI: 10.1016/j.seppur.2022.122427

[80] Liu R, Geng Y, Tian Z, Wang N, Wang M, Zhang G, et al. Extraction of platinum(IV) by hydrophobic deep eutectic solvents based on trioctylphosphine oxide. Hydrometallurgy. 2021;**199**:105521. DOI: 10.1016/j.hydromet.2020.105521

[81] Vargas SJR, Peréz-Sánchez G, Schaeffer N, Coutinho JAP. Solvent extraction in extended hydrogen bonded fluids—Separation of Pt(IV) from Pd(II) using TOPO-based type V DES. Green Chemistry. 2021;**23**:4540-4550. DOI: 10.1039/D1GC00829C

[82] Schaeffer N, Conceição JHF, Martins MAR, Neves MC, Pérez-Sánchez G, Gomes JRB, et al. Non-ionic hydrophobic eutectics—Versatile solvents for tailored metal separation and valorization. Green Chemistry. 2020;**22**:2810-2820. DOI: 10.1039/D0GC00793E

[83] Liu R, Hao J, Wang Y, Meng Y, Yang Y. A separation strategy of Au(III), Pd(II) and Pt(IV) based on hydrophobic deep eutectic solvent from hydrochloric acid media. Journal of Molecular Liquids. 2022;**365**:120200. DOI: 10.1016/j.molliq.2022.120200

Chapter 4

Extraction of Boron from Tourmaline Ore: Mechanism of Thermal Analysis of the Schorl

Sneha Dandekar, Kavita Pande and Dilip Peshwe

Abstract

Tourmaline is the chief boron-bearing mineral on the earth and is present in an excess amount in the crustal rocks. However, schorl is an iron-rich alkali that forms a solid solution with the magnesium-rich, alkali tourmaline, dravite. In this work, tourmaline (schorl variety) was treated along with soda ash, and its behavior was analyzed using electron probe microanalysis (EPMA), x-ray diffraction (XRD), scanning electron microscope, and energy dispersive spectrometer (SEM-EDS) analyses, thermogravimetric analysis (TGA), differential thermal analysis (DTA), in order to annotate the soda-ash activation of boron within the tourmaline ore. To extract boron from the sample, sodium carbonate powder was mixed with the schorl in 20% of the total weight of schorl powder. When the sample was treated with sodium carbonate, the sodium gets combined with the boron to form sodium borate at 566°C along with aegirine and aluminum oxides. This sodium borate can be treated with hydrochloric acid to get boron-oxide along with NaCl.

Keywords: boron, tourmaline, sodium carbonate, sodium borate, boron-oxide, extraction

1. Introduction

Tourmaline is an important and complex boron-bearing mineral on the earth and is present in an excess amount in the crustal rocks. It is not a single mineral, but a group of isomorphs minerals with identical crystal lattices. The general formula of the tourmaline group is very complex: X1Y3Al6B3Si4O27 (OH) $_4$ where X = Na, K, Ca and Y = Li, Fe, Mg, Mn, Al, Cr, Ti; some of the OH– ions are generally substituted by F. The boron concentration in the tourmaline has been found 3.40 ± 1% within the normal range (2.5–3.8%) reported worldwide [1]. Whereas, schorl is an iron-rich alkali tourmaline that forms a solid solution with the magnesium-rich, alkali tourmaline, dravite. It is reported that, economically sustainable deposits of borax have not been established in India so far. The only deposit of little economic importance is reported from Puga Valley in Leh district, Jammu & Kashmir. As per National Mineral Inventory data, based on the UNFC system, the total reserves/resources of borax as of 1.4.2015, have been estimated at 74,204 tons in Jammu & Kashmir. Occurrences are also reported from Surendranagar district, Gujarat, and Jaipur district, Rajasthan [2].

Boron does not occur in free state in nature. It occurs mainly in the form of the salts of boric acid. It mainly occurs in the form of boric acid (H_3BO_3) and borax ($Na_2B_4O_7.10H_2O$). Element boron and its compounds are widely used in the industrial and agricultural sectors due to their properties such as high hardness, wear resistance, high strength, fire inhibitor, heat resistance, high strength, wear resistance, and catalytic performance. In general, boron–iron separation and dissolution activity of boron-bearing minerals in alkaline liquor are the two key issues in the utilization of tourmaline ore, governing the boron recovery as well as operating cost [3]. On a large scale, boron is extracted from its minerals, borax $Na_2B_4O_7$ or colemanite $Ca_2B_6O_{11}$. The latter is first converted to borax by boiling with a solution of sodium carbonate in the requisite proportion. In the present study, attempt has been made to extract boron by thermal analysis method.

2. Materials and methodology

Tourmaline (Schorl) was obtained from the Kyanite, Sillimanite mine of Girola area, Sakoli tehsil, Bhandara. The samples were collected, crushed, ground, and separated to 80% passing through 200 sieve size (0.074 mm). The XRD analysis of the sample was completed on Panalytical X'Pert Pro (model-PW 3040/60) diffractometer with Cu Kα radiation (λ = 1.54 Å) produced at a voltage of 45 kV and current of 40 mA. Scanning was done at the 2θ angle of 10 to 100° with a scan step size and time per step of 0.01° and 15 seconds, respectively. The surface morphology was examined using a Scanning Electron microscope (SEM-JEOL 6830A). Before the study of surface morphology, to make the material electron-conducting it was coated with a thin platinum coat using an auto sputter (JOEL-JFC 1600 auto fine coater). Whereas, the elemental composition was studied using Energy Dispersive Spectroscopy (EDS).

Electron-probe microanalysis (EPMA) of the sample was executed at the National Centre of Excellence in Geoscience Research, GSI, Bangalore using a CAMECA SX-100 electron microprobe analyzer. The spectra were collected for each sample with a wavelength-dispersive spectrometer (WDS) and with WDS 1 (TAP crystal), WDS 4 (TAP crystal), WDS 3 (LPET crystal), WDS 2 (PET crystal), and WDS 5 (LIF crystal) spectrometers. The spectra were collected using column condition of an acceleration voltage of 15 keV, beam current of 15 nA, and beam size of 1 μm. Calibration, quantification, and overlap correction were executed using CAMECA SX- 100 Peak Sight-Geo Quanta software package.

However, to extract boron from the sample, sodium carbonate powder was mixed with the schorl in 5, 10, 15, and 20% of the total weight of schorl powder. The virgin, as well as prepared samples, were roasted to 1000°C using a tubular furnace [4], and analyses were done. However, suitable boron reach compound was found to be in 20% proportion of sodium carbonate powder, hence further analysis was done of this composition by leaching out with HCl (**Figure 1**). In order to achieve the boron in pure form, acid leaching was done multiple times. However, by further synthesis iron can be removed from the compound, followed by the recovery of aluminum.

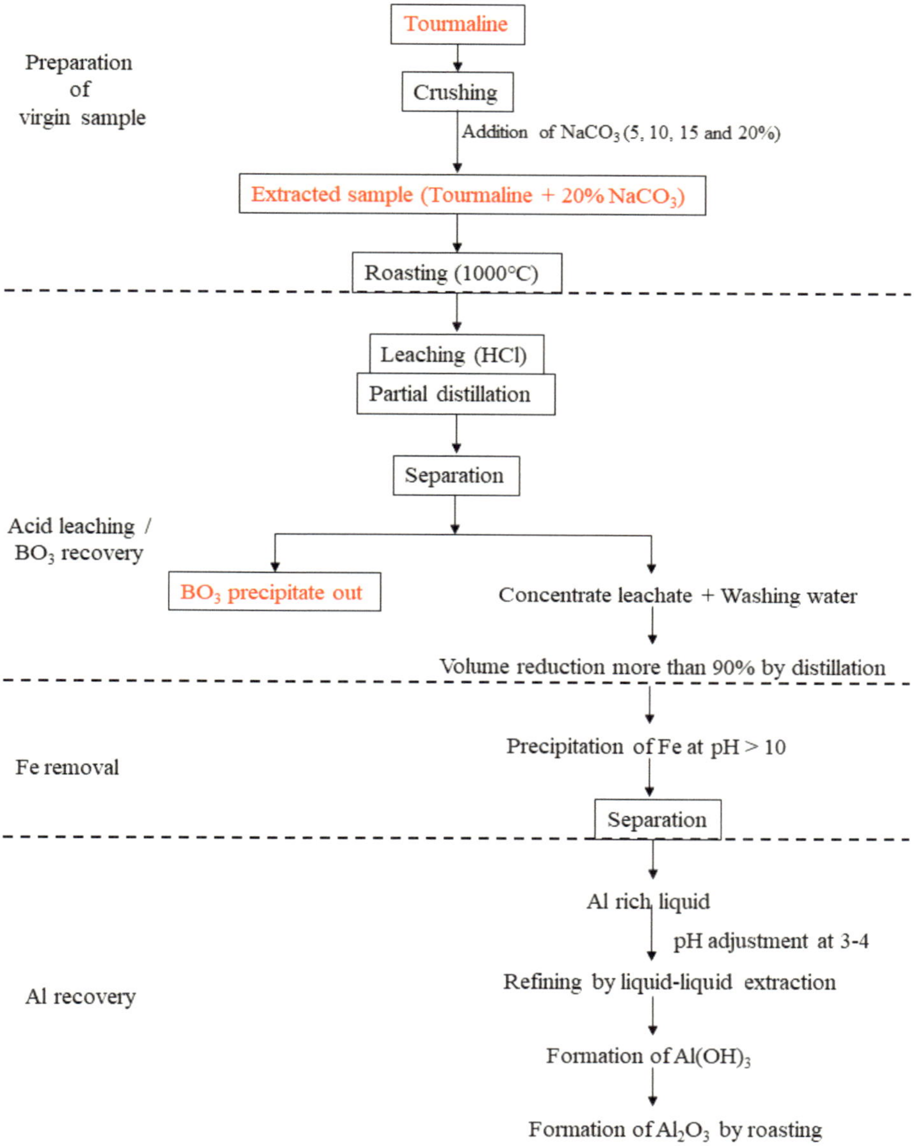

Figure 1.
Flow chart showing the extraction process.

3. Results

3.1 EPMA

The EPMA study was carried out for the extracted sample. From the EPMA study of the sample (**Table 1**), the mineral shows the highest content of Al_2O_3 of 25.88 to 61.9 wt. %, SiO_2 ranges from 30.01 to 44.89 wt. %, TiO_2 ranges from 0 to 1.82 wt. %, MnO content is 0 to 9.59 wt. %, MgO ranges from 0 to 14.07 wt. %, CaO ranges from

	1	2	3	4	5	6	7	8	9	10
SiO_2	30.74	32.73	33.58	32.92	33.13	34.04	36.5	37.22	37.68	31.43
TiO_2	0	2.87	1.63	0.54	2.19	0.39	0	0.07	0.03	0.02
Al_2O_3	1.4	16.26	30.62	30.7	23.38	27.33	40.07	38.03	33.14	46.53
Fe_2O_3	43.89	14.78	0.86	12.37	5.44	7.63	0	0	0	0
FeO	2.69	11.24	12.65	5.98	8.66	4.91	0.22	0.1	1.95	0.05
MnO	0	0.1	0.06	0.11	0.07	0	3.07	4.06	0	0.02
MgO	6.45	4.66	2.69	0.16	7.8	7.34	0	0.18	9.71	0
CaO	0	0.04	0	0.2	3.3	0.99	0.2	2.63	0.03	1.74
Na_2O	2.12	2.71	2.84	2.49	1.16	2.35	2.15	1.4	1.42	1.33
K_2O	1.04	0.19	0.06	0.07	0.05	0	0	0.02	0	0
B_2O_3	9	9	10	10	10	10.83	11.56	0	0	16.2
H2O	2.56	3	3	1	3	3.18	3.13	3.17	0	3.25
Total	100.1	98.16	98.571	97.594	98.66	99	100.832	101.859	100.434	101.849
Unit formula normalized to 31 anions										
Si	5.878	5.956	5.785	5.644	5.771	5.792	5.877	5.941	5.995	5.149
Ti	0	0.393	0.211	0.07	0.287	0.05	0	0.008	0.004	0.002
Al	0.316	3.487	6.217	6.204	4.8	5.481	7.604	7.154	6.214	8.984
Fe3+	6.315	2.024	0.111	1.596	0.713	0.977	0	0	0	0
Fe2+	0.43	1.711	1.823	0.857	1.262	0.699	0.03	0.013	0.259	0.007
Mn	0	0.015	0.009	0.016	0.01	0	0.419	0.549	0	0.003
Mg	1.839	1.264	0.691	0.041	2.026	1.862	0	0.043	2.303	0
Ca	0	0.008	0	0.037	0.616	0.18	0.035	0.45	0.005	0.305
Na	0.786	0.956	0.949	0.828	0.392	0.775	0.671	0.433	0.438	0.422

	1	**2**	**3**	**4**	**5**	**6**	**7**	**8**	**9**	**10**
Li	0	0.004	0.022	0	0	0.003	1.041	1.194	0	0.369
B	3.027	3	3.081	3.001	3.121	3.181	3.213	3	3	4.581
Name	**Schorl**	**Schorl**	**Oxy-Schorl**	**Dumortierite**	**Dumortierite**	**Dravite**	**Schorl**	**Schorl**	**Dumortierite**	**Schorl**

Table 1.
Showing the composition of virgin sample.

0 to 6.92%, Na_2O content is 0 to 0.95 wt. %, and K_2O ranges from 0 to 0.93 wt. %. The mineral shows a high percentage of Al and considerable x-site vacancy, an intermediary between schorl, dumortierite, and dravite.

However, the EPMA analysis of the extracted sample (**Table 2**) shows the highest range of SiO_2 from 41.75 to 42.8 wt. %, MgO ranges from 14.3 to 15.27 wt. %, CaO 11.15 to 11.19 wt. %, and Al_2O_3 9.33 to 9.49 wt. %. The content shows the presence of aegirine mineral [5].

3.2 XRD

In order to explain borate formation, an XRD analysis of the samples was done. **Figure 2** shows the XRD patterns of the untreated and treated samples. In **Figure 2a**, peak of schorl ($NaFe^{2+}{}_3Al_6$ (Si_6O_{18}) $(BO_3)_3(OH)$ $3(OH))$ [6], dumortierite (Al_7 BO_3 $(SiO_4)3O_3$) [7] and SiO_2 [8] was found. However, after the extraction (**Figure 2b**), the identified peaks correspond to sodium borate ($Na_2B_4O_7$) [9], aegirine ($NaFeSi_2O_5$) [10], and Al_2CO_3 [11]. However, after multiple times acid leaching of the extracted sample, boron oxide was recovered (**Figure 2c**).

3.3 SEM

The surface morphology of the sample has been shown in **Figure 3**. The formation of orthorhombic-shaped dumortierite crystals [12] has been observed in SEM (**Figure 3a**). However, schorl is found in well-crystallized hexagonal-shaped along with kyanite (**Figure 3b**). The dumortierite crystals are formed after the disintegration of the tourmaline grain. However, in the extracted sample, the formation of sodium borate along with aegirine was observed. From this, it can be stated that, at a particular temperature, schorl along with sodium carbonate can form sodium borate [13] (**Figure 3c**) and aegirine ((**Figure 3d**). The acid leaching helps in the extraction of boron oxide from the mineral (**Figure 3e** and **f**), as the sodium borate reacts with HCl forming salts and boron-oxide. The Fe and Al residue gets dissolved in the solute, which can be extracted by further synthesis.

3.4 TGDTA

Three major weight losses are seen in **Figure 4a**. The first weight loss up to 85°C accounts for the dehydration of the untreated sample, which produced an endothermic peak. The second weight loss and an exothermic peak are seen at 320°C. This exothermic peak corresponded to the crystallization of $FeAlO_3$, oxidation of Fe^{3+}, and oxidation of Fe^{2+} to Fe^{3+}. Further, an increase in temperature also triggered the decomposition of the anhydrous salt, contributing to the second weight loss. The third weight loss up to 857°C accounts for the decomposition of residual nitrate and the phase transformation of tourmaline. Therefore, the temperature of 650°C was selected as the heat-treatment temperature to protect the crystalline structure and improve the far-infrared emission of the tourmaline [14].

However, the TGDTA peak of a treated sample (**Figure 4b**). First exothermic peak at 566°C was due to the crystallization of anhydrous sodium borate (Na_2 O $(B_2O_3)_2$) from the amorphous phase [9, 15]. Another endothermic peak observed at 742°C was due to the melting of the crystalline anhydrous sodium borate phase.

DOI: http://dx.doi.org/10.5772/intechopen.111595

Oxides	**1**	**2**	**3**	**4**	**5**	**6**	**7**	**8**
SiO_2	42.63	42.83	42.8	41.75	42.83	42.8	42.63	41.75
TiO_2	1.21	1.12	1.1	1.23	1.12	1.1	1.21	1.23
Al_2O_3	9.49	9.39	9.33	9.39	9.39	9.33	9.49	9.39
Fe_2O_3	7.24	7.15	7.2	7.24	7.15	7.2	7.24	7.24
FeO	7.06	7.1	7.13	7.1	7.1	7.13	7.06	7.1
MgO	14.32	14.27	14.3	15.27	14.27	14.3	14.32	15.27
CaO	11.19	11.15	11.17	11.15	11.15	11.17	11.19	11.15
Na_2O	1.35	1.25	1.29	1.25	1.25	1.29	1.35	1.25
K_2O	0.11	0.16	0.19	0.16	0.16	0.19	0.11	0.16
H_2O (+)	4.87	4.69	4.7	4.89	4.69	4.7	4.87	4.89
H_2O (−)	0.12	0.17	0.2	0.15	0.17	0.2	0.12	0.15
Total	99.59	99.28	99.41	99.58	99.28	99.41	99.59	99.58
Fe_2O_3/FeO	1.03	1	1	1.01	1	1	1.03	1.01
Fe_2O_3/Al_2O_3	0.76	0.76	0.77	0.77	0.76	0.77	0.76	0.77
Name	**Aegirine**	**Aegirine**	**Aegirine**	**Aegirine**	**Aegirine**	**Aegirine**	**Aegirine**	**Aegirine**

Table 2.
Showing the composition of extracted sample.

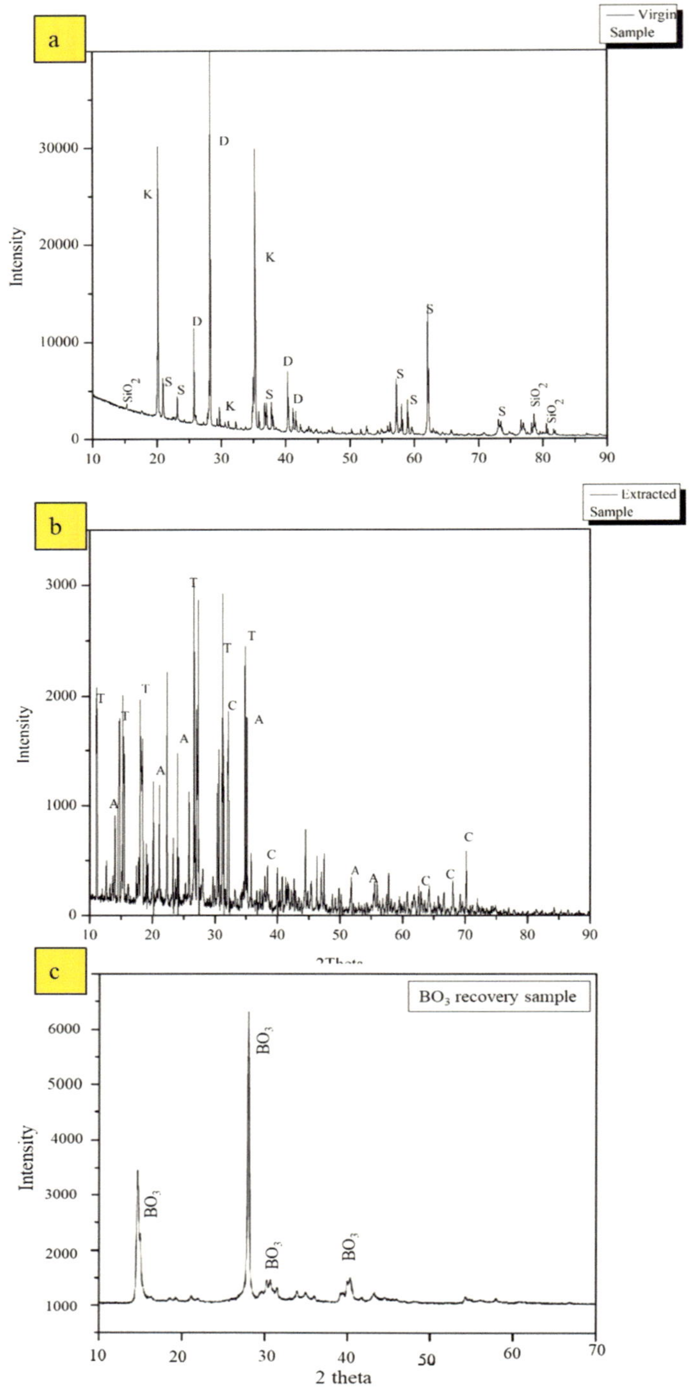

Figure 2.
XRD graph of (a) virgin (K- Kyanite, D-Dumortierite, S-Schorl) (b) extracted sample (A-Aegirine, T-sodium borate, C- Aluminum oxide) (c) boron recovery sample.

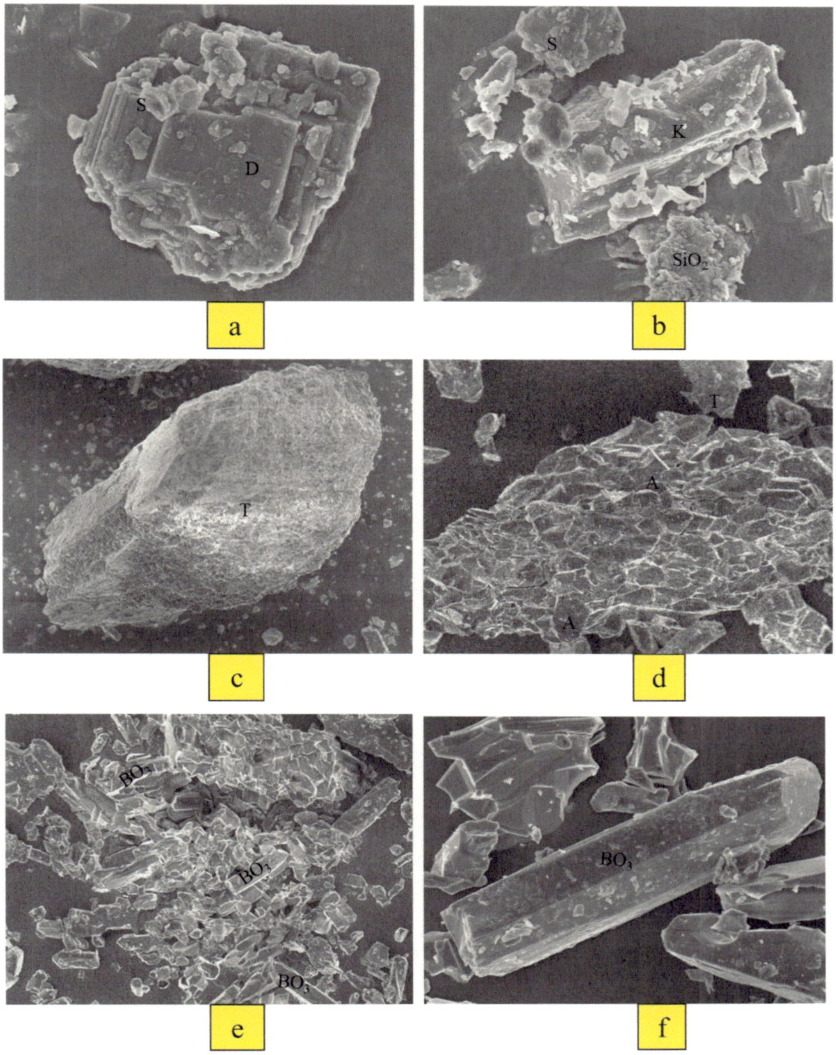

Figure 3.
SEM images of (a, b) virgin sample (c, d) extracted sample, where S- schorl, D- dumortierite, T- sodium borate, and A- aegirine, and (e, f) extracted BO_3.

4. Discussion

In order to extract boron from the tourmaline sample, it was treated with 20% of sodium carbonate power, in order to form a boron compound that will ease the element to get out of the complex tourmaline structure. On heating, the virgin, as well as extracted samples up to 1000°C following conclusions, were observed:

1. On heating the virgin sample, iron-aluminum oxides were formed at 320°C, and at the higher temperature, a transformation of tourmaline was observed. Hence, no liberation of the boron compound was found in this case.

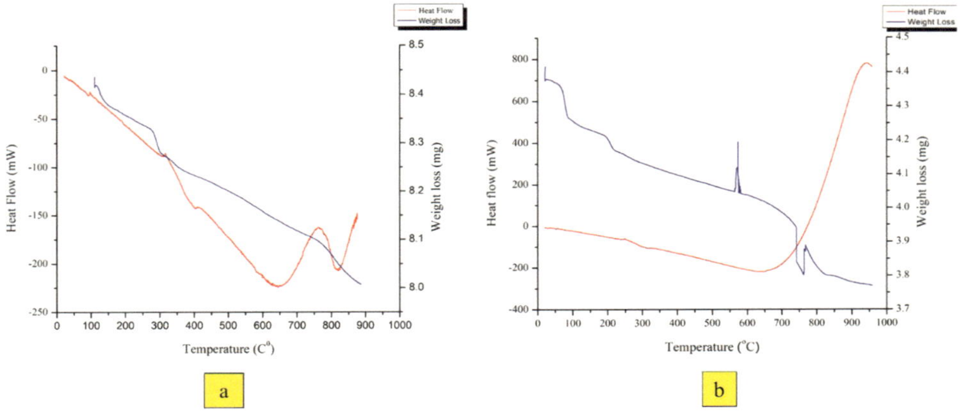

Figure 4.
TGDTA graph of (a) virgin (b) extracted sample.

2. Whereas, for the extracted sample, following reactions 1 and 2 were observed, which can state the formation of sodium borate and boron oxides, respectively.

$$Fe^{2+}{}_3Al_6\left(Si_6O_{18}\right)\left(BO_3\right)_3\left(OH\right)_3\left(OH\right)+2\mathrm{Na}\left(CO_3\right)_3 \rightarrow Na_2\mathrm{B}_4\mathrm{O}_5\cdot 8\,\mathrm{H}_2\mathrm{O}+\mathrm{NaFeSi}_2\mathrm{O}_6+\mathrm{AlCO}_3 \quad (1)$$

$$Na_2\mathrm{B}_4\mathrm{O}_5\cdot 8\,\mathrm{H}_2\mathrm{O}+\mathrm{NaFeSi}_2\mathrm{O}_6+\mathrm{AlCO}_3+4\,HCl \rightarrow 4\,\mathrm{B}\left(OH\right)_3+4\,\mathrm{NaCl}+8\mathrm{H}_2\mathrm{O}+\mathrm{FeCO}_3\uparrow+Al_2\mathrm{O}_3\cdot SiO_2+4\,\mathrm{H} \quad (2)$$

As, when the sample was treated with sodium carbonate, the sodium gets combined with the boron to form sodium borate at 566°C along with aegirine and aluminum carbonate (Eq. (1)). Moreover, the sodium borate when treated with hydrochloric acid forms boron-oxide along with NaCl (Eq. (2)). However, further synthesis process can be followed to remove the Fe and Al recovery from the leachate.

5. Concluding remarks

In order to extract boron from the tourmaline sample, it was treated with 20% $NaCO_3$ and heated up to 1000°C. It was found that the sodium from the $NaCO_3$ powder gets combined with the boron and forms sodium borate. However, the boron-oxide can be extracted by treating the sodium borate with hydrochloric acid. Extraction of boron by this method is cost-effective and can fulfill the need for boron elements in various industries.

Author details

Sneha Dandekar[1*], Kavita Pande[2] and Dilip Peshwe[1]

1 Department of Metallurgical and Materials Engineering, Visvesvaraya National Institute of Technology, Nagpur, India

2 Mataverse Vision Pvt. Ltd., Nagpur, India

*Address all correspondence to: snehandandekar@gmail.com

References

[1] Qureshi AA, Akram M, Khan MA, Khattak NU, Qureshi IE, Khan HA. Boron determination in tourmaline by neutron induced radiography. Radiation Measurements. 2001;**34**(1-6):345-348. DOI: 10.1016/S1350-4487(01)00183-4

[2] Bhavan I. Indian Minerals Yearbook 2020 59th Edition Boron Minerals (Advance Release) Government of India Ministry of Mines Indian Bureau of Mines. 2022. www.ibm.gov.in

[3] Chen J, Liu S, Zhang X. Study of carbonate alkaline leaching boron-rich slag for borax preparation. Journal of Northeastern University (Natural Science). 1996;**17**(5):508-511

[4] Zou C, Tang Z, Xie W, Fu H, Kuang J, Deng Y, et al. Effects of sodium roasting on the leaching rate of boron-bearing tailings and its mechanism analysis. Royal Society Open Science. 2018;**5**(8):172342. DOI: 10.1098/rsos.172342

[5] Watanabe T. Aegirine-augite and sodic augite from the Sambagawa and the Chichibu belts in the Oshika District, Central Japan, with special reference to Na-Metasomatism, journal of the Faculty of Science, Hokkaido University. Geology and Mineralogy. 1975;**16**(4):519-532

[6] Bloodaxe ES, Hughes JM, Dyar MD, Grew ES, Guidotti CV. Linking structure and chemistry in the schorl-dravite series. American Mineralogist. 1999;**84**:922-928

[7] Moore PB, Araki T. Dumortierite, $Si_3 B[Al_{6.75\ 0.25}O_{17.25}(OH)_{0.75}]$: A detailed structure analysis. Neues Jahrbuch für Mineralogie, Abhandlungen. 1978;**132**:231-241

[8] Joni IM, Nulhakim L, Vanitha M, Panatarani C. Characteristics of crystalline silica (SiO2) particles prepared by simple solution method using sodium silicate (Na2SiO3) precursor. Journal of Physics: Conference Series. 2018;**1080**(1):012006. DOI: 10.1088/1742-6596/1080/1/012006

[9] Lim KY, Kim YW, Song IH. Porous sodium borate-bonded SiC ceramics. Ceramics International. 2013;**39**(6):6827-6834. DOI: 10.1016/j.ceramint.2013.02.014

[10] Sebakhy KO, Vitale G, Pereira-Almao P. Dispersed Ni-doped Aegirine Nanocatalysts for the selective hydrogenation of Olefinic molecules. ACS Applied Nano Materials. 2018;**1**(11):6269-6280. DOI: 10.1021/acsanm.8b01472

[11] Ansari SA, Husain Q. Immobilization of Kluyveromyces lactis β galactosidase on concanavalin A layered aluminium oxide nanoparticles - its future aspects in biosensor applications. Journal of Molecular Catalysis B: Enzymatic. 2011;**70**(3-4):119-126. DOI: 10.1016/j.molcatb.2011.02.016

[12] Sabzehei M. Dumortierite from Iran: A first record. Mineralogical Magazine. 1971;**38**(296):526-527. DOI: 10.1180/minmag.1971.038.296.19

[13] Guo J, Chen X, Zhang Y. Improving the mechanical and electrical properties of ceramizable silicone rubber/halloysite composites and their ceramic residues by incorporation of different borates. Polymers. 2018;**10**(4):388. DOI: 10.3390/polym10040388

[14] Zhu D, Wang L. Effect of tourmaline modified with La-doped nano-CeO2 on consumption of liquefied petroleum gas Antibacterial Behavior of Ag-Ce-TiO

2 Composite Films in the Different Irradiations View project. Journal of Rare Earths. 2007;**25**:150 https://www.researchgate.net/publication/295727316

[15] Hawthorne FC, Henry DJ. Classification of the minerals of the tourmaline group. European Journal of Mineralogy. 1999;**11**:201-215

Section 3

Cast Iron Smelting

Chapter 5

Evaluation of Metallurgical Quality of Cast Iron Using Quality Criteria

Peter Futas and Alena Pribulova

Abstract

The metallurgical quality of the produced cast iron is related to its chemical composition (mainly the content of C, Si, Mn, P, and S), or other monitored elements—alloying elements (Cr, Ni, Cu, ...), in some cases showing elements (Pb, Sn, As, Sb, ...). The chemical composition of cast iron is determined by the degree of saturation (Sc) or carbon equivalent (CE). Other factors influencing the quality of cast iron are the metallurgical conditions of production (melting and treatment) of cast iron and the rate of solidification in the mold. The mechanical properties of cast iron (Rm, HB, and E0) are closely related to its chemical composition. In addition to this common evaluation of cast iron, other quality criteria of gray cast iron are also used in practice. This is a comparison of the mechanical properties of the produced gray cast iron with the optimal values determined for the same degree of saturation (Sc). This chapter concerns assessments of the metallurgical quality of gray cast iron and the results of operational melting of synthetic gray cast iron with different charge compositions in the Slovak Foundry and its analysis.

Keywords: metallurgy, quality criteria, cast iron, properties, steel scrap

1. Introduction

Cast irons are alloys of iron, carbon, silicon, manganese, and other elements, while carbon is excluded in the form of graphite or is bound as carbide Fe_3C or carbide of another element. The carbon content exceeds the value of the maximum solubility of carbon in austenite, that is, C > 2.06% without the influence of other elements. Crystallizes according to the stable Fe–C or metastable Fe–Fe_3C diagram, or during solidification and cooling using both systems (**Figure 1**) [1].

Gray cast iron is called gray because of the gray color on the fracture surface. It contains 1.5–4.3% C and 0.3–5% Si plus Mn, S, and P. It is brittle with low tensile strength but has good foundry properties.

The structure of cast iron is characterized by randomly oriented graphite flakes (**Figure 2**) [2]. The more sharp-edged the graphite formations are, the greater the notch effect, which reduces the plasticity of cast iron and increases brittleness.

The structure of gray cast iron usually has three phases: ferrite, pearlite, or martensite. The carbon content in the metal matrix exceeds 1%. It contains graphite in the form of three-dimensional structures, which have the shape of flakes on the metallographic surface. Their distribution and shape largely influence the properties of the

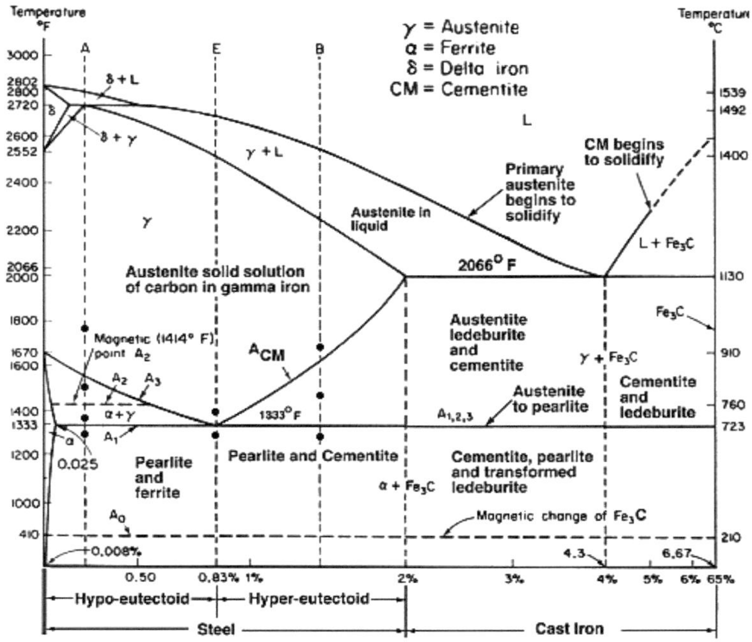

Figure 1.
Fe-Fe3C phase diagram [1].

Figure 2.
Microstructure of gray cast iron [2].

material. The best gray cast irons have a metal matrix with graphite flakes of varying size and uniform distribution.

The chemical composition of gray cast iron should be chosen to ensure the following:

- Desired shape and distribution of graphite
- Carbide-free structure
- Required metal matrix

The structure of the metal matrix affects the properties as follows:

- Ferrite matrix—low strength
- Perlite matrix—higher strength
- Martensitic matrix—high strength and higher hardness when high-carbon cast irons are quenched.

Besides carbon, silicon is the most important element in cast iron. During solidification, it significantly supports graphitization, and during the transformation of austenite, it supports the formation of ferrite.

Gray cast iron is a widely used material in the automotive industry for engine blocks, brake discs, brake drums, and covers. It has good machinability with excellent wear resistance and excellent vibration damping. Thanks to its physical properties, it is preferred by designers and enables the production of castings with excellent specific properties, especially heat and fire resistant, abrasion resistant, and castings with special physical properties [3]. Its disadvantage is its fragility and relatively large dispersion of properties, especially mechanical ones, even with the same chemical composition. Solving this phenomenon requires excellent metallurgical and technological knowledge [4].

The worldwide trend of gray iron shares 54.4% (51,190,987 tons in 2021) [5].

2. Quality evaluation of gray iron

The quality of gray cast iron, usually expressed by values of mechanical properties (Rm, HB, and E_0), is closely related to its chemical composition (content of C, Si, Mn, P, and S) eventually other monitored elements—alloying elements (Cr, Ni, Cu, …), in some cases, pollutants (Pb, Sn, As, Sb, …) [6].

From a chemical point of view, gray cast iron is evaluated from the three main categories [7]:

1. *The main elements:* These are carbon, silicon, and iron. Gray cast irons typically contain 3.0–3.5% carbon, with silicon levels varying from 1.8 to 2.4%. Although increasing the carbon and silicon contents improves the graphitization potential and therefore decreases the chilling tendency, the strength is adversely affected. This is due to ferrite promotion and the coarsening of pearlite.

2. *The minor elements:* Phosphorus and the two related elements, manganese and sulfur. Phosphorus is found in all gray irons, and although rarely added intentionally, it does increase the fluidity of iron to some extent. Thus, most castings are produced with 0.02–0.10% P. Sulfur plays a significant role in the nucleation of graphite in gray iron, with optimum benefit at 0.05–0.12% sulfur levels. It is also important to note that sulfur levels need to be balanced with manganese to promote the formation of manganese sulfides. Typically, it can be as low as 0.1% for ferritic irons and as high as 1.2% for pearlitic irons because manganese is a strong pearlite promoter.

The optimum ratio between manganese and sulfur for a FeS-free structure and maximum amount of ferrite is [8]:

$$\%Mn = 1.7x(\%S) + 0.15$$

Other minor elements, such as aluminum, antimony, arsenic, bismuth, lead, magnesium, cerium, and calcium, can significantly alter both the graphite morphology and the microstructure of the matrix [9, 10].

3. *The trace elements:* Many other elements are utilized in limited amounts to affect the nature and properties of gray iron. Although some are not intentional, they do have a measurable effect on the gray cast iron. Some promote pearlite, such as tin, while others compact graphite and increase strength, such as nitrogen [11, 12].

The chemical composition is simply determined by the carbon equivalent (CE) or degree of saturation (Sc) [13].

2.1 Carbon equivalent

It takes the influence of the accompanying elements equal to the effect of the carbon into account and provides information on the composition relative to the eutectic composition in analogy to the degree of saturation, that is, it expresses the equivalent carbon content with respect to the iron-carbon binary system [6]:

$$CE = C + \frac{Si + P}{3} \tag{1}$$

In eutectic compositions, CE equals 4.3%; the following relationships exist between the degree of saturation and the carbon equivalent CE, mainly used in the Anglo-American countries [6]:

$$Sc = \frac{C}{4.23 - CE + C} \tag{2}$$

$$CE = 4.3 + C\left(1 - \frac{1}{Sc}\right) \tag{3}$$

In all formulas, C is to be replaced with the total carbon content.

2.2 Degree of saturation

It indicates the ratio between the total carbon content of the melt and the carbon content of the eutectic composition. According to Eq. (4), it is under consideration of the influence of accompanying elements on the shifting of the eutectic point [6]:

$$Sc = \frac{C}{4.26 - \frac{1}{3}(Si + P)} \tag{4}$$

If other accompanying elements in the iron or alloy components are also considered, this results in the following exact formula Eq. (5) [6]:

$$Sc = \frac{C}{4.26 - 0.31.Si - 0.27.P - 0.4.S - 0.074.Cu + 0.063.Cr + 0.02.Mn} \tag{5}$$

A degree of saturation Sc = 1.0 means that the iron corresponds exactly to the eutectic composition.

Degrees of saturation above 1.0 have hypereutectic cast iron and thus result in the structural formations of cast iron to be expected under consideration of the relevant wall thicknesses. In the chemical composition of flake graphite cast iron, carbon has the strongest influence on the strength.

The lower the carbon content and the degree of saturation, the greater the strength because less graphite and more primary dendrites are present in the structure. Manganese, phosphorous, and sulfur in the usual contents have only a minor influence on tensile strength. Inoculation treatment results in an increase in the number of eutectic cells and thus in greater strength. The so-called Sipp diagram (**Figure 3**) [14], illustrates the correlations with regard to the material structure and the Heller-Jungbluth diagram (**Figure 4**) [14], those with regard to the tensile strength.

The degree of saturation can also be used for the approximate calculation of the tensile strength of gray cast iron depending on the wall thickness. According to Eq. (6) [6]:

$$\text{Normal tensile strength Rm (MPa)} = 1000\text{–}800\ \text{Sc (for the standard 30 mm test bar)} \quad (6)$$

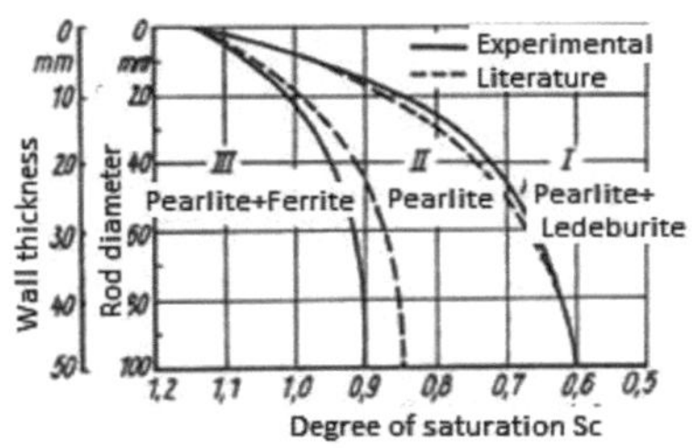

Figure 3.
Relation between the degree of saturation and the basic structure [14].

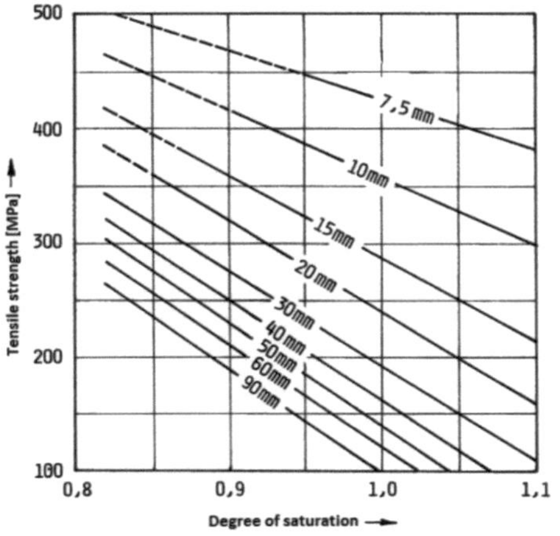

Figure 4.
Relation between the degree of saturation and the tensile strength of flake graphite cast iron [14].

The quotient of measured tensile strength and calculated normal tensile strength indicate the so-called degree of normality, an important quality parameter of the flake graphite cast iron.

There is a correlation between the degree of saturation and the carbon equivalent.

For most applications, a gray, pearlitic cast iron with a degree of saturation of approximately 0.85 to 0.95 for wall thicknesses of around 15 mm is preferred, with a carbon content of approximately 3.1 to 3.3% and a silicone content of 1.5 to 1.8%. As a general rule, the Si/C proportion for a given degree of saturation should be adjusted to a possibly high value.

They are used in quality criteria of gray iron in engineering practice, in addition to ordinary quality evaluation of cast iron, that is, determination of Rm and HB, eventually chemical composition or other required properties. This is a comparison of the mechanical properties of gray iron produced with optimal values determined for the same degree of saturation.

2.3 Regression equations for calculation of mechanical properties

Statistical methods (regression equations with multiple correlations) determine the quantitative relationship among the chemical compositions (basic, if necessary, accompanied by other observed items), or the degree of saturation or carbon equivalent CE and strength and toughness of gray cast iron, in some cases modulus of elasticity E_0.

The most used equations to calculate the mechanical properties of pearlitic gray iron from its chemical composition are [6, 14]:

$$\text{Rm} = 786.5–150\%\text{C}–47\%\text{Si} + 45\%\text{Mn} + 219\%\text{S}; \text{variation s} = 25\ \text{MPa} \quad (7)$$

$$\text{HB} = 444–71.8\%\text{C}–13.9\%\text{Si} + 21\%\text{Mn} + 170\%\text{S}; \text{variation s} = 12.46 \quad (8)$$

$$E_0 = 313.175–449.014\%\text{C} - 14082\%\text{Si}; \text{variation s} = 6760\ \text{MPa} \quad (9)$$

The most appropriate procedure is to calculate the regression equation for its own foundry (i.e., for the specific conditions).

2.4 The quality criteria

The degree of maturity of cast iron RG specifies the level of quality of gray iron production compared with the optimal, determined by calculation, value of Sc, respectively, by correlation Eq. (6):

$$RG = \frac{Rm_{measured}}{Rm_{calculated}} \bullet 100\% \quad (10)$$

where $Rm_{calculated}$ = 1000–800 Sc

It also determines *the relative hardness RH* [6]:

$$RH = \frac{HB_{measured}}{HB_{calculated}} \bullet 100\% \quad (11)$$

where $HB_{calculated}$ = 100 + 0.44 $Rm_{measured}$.

A value above 100% means a high quality of gray iron.

The quality number GZ or quality factor is obtained by dividing the RG/RH or by Rm/HB—measured values. Qualitative gray iron has high strength at low hardness.

3. Evaluation of quality of gray iron in a particular foundry

In the operational, conditions of foundry in Slovakia have analyzed 122 melts of gray iron EN-GJL-250 with different charges. The steel scrap rate was 77.82% and the return material was 22.18%. In the foundry, also one melt with a 100% rate of the steel scrap in the charge was realized (synthetic gray iron).

These melts were realized in two low-frequency induction furnaces "Siemens" with the following parameters:

- the nominal content of the furnace: 6 tons
- melting furnace power: 5350 kg/h
- nominal frequency: 50 Hz

The cast iron was by the bar gauge treated FeSi75. The temperature of the superheated melt was between 1420 and 1440°C.

Next samples were a cast from the melts:

- samples for chemical analysis
- test bars with a diameter of 30 mm (tensile strength—Rm, Brinell Hardness HB)

The samples for metallographic analysis were taken from the test bars and prepared in a standard manner.

The hardness was measured with the durometer HPO 3000 (terms: 10/3000/10), that is, the diameter of the ball was 10 mm, strength duration was 3000 N, and period of load was 10 sec.

Tensile strength was measured on test bars (Ø30 mm) on the universal crackle machine ZWICK.

The statistical results of the chemical composition and measured mechanical properties are documented in **Table 1**.

The stress results of gray iron (EN1561) show small dispersion of Rm (260–285 MPa), the average value of Rm = 274 MPa, which confirms the uniformity of production. The decrease of Rm by an increase of Sc is slight (**Figure 5**).

For the melt with 100% steel scrap, the charge hardness that was greater at approximately 45% (HB 293) was recorded. But on the other side, the gray iron had lower tensile strength (Rm = 226 MPa) against average values from other melts.

More than half of the melts were produced with a lower Sc than the specified EN (Sc min. 0.87). The average value of Sc = 0.850 is below this value too. It is possible to expect the opposite result in the production of synthetic gray iron (100% steel scrap).

The average value HB = 201 is optimal for this brand of gray iron even though the dispersion of the values HB (183–225) is relatively high in a whole range of the chemical composition (Sc = 0.798–0.934) (**Figure 6**).

On the base of multiple regressions [15], dependence equations of tensile strength Rm and Brinell Hardness HB on C, Mn, Si, P, S, and Cr were calculated.

Furnace charge average n = 122	C	Mn	Si	S	P	Cr	Sc	CE	Rm [MPa]	HB
	(%)									
Return material	3.211	0.826	1.575	0.0624	0.136	0.116	0.850	3.664	274	201
22.18% steel	0.074	0.0711	0.096	0.0437	0.05	0.034	0.1114	0.4781	6.0415	11.23
scrap 77.82%	3.0	0.66	1.19	0.024	0.06	0.04	0.798	3.456	260	183
steel	3.4	0.98	1.81	0.094	0.27	0.18	0.934	3.988	285	225
scrap 100%	3.28	0.78	1.79	0.063	0.18	0.09	0.901	—	226	293

NOTE: the first line—*arithmetic average,* the second line—*standard deviation,* the third line—*the lowest value of the file,* the fourth line—*the highest value of the file, and n—number of melts.*

Table 1.
The statistical results of the chemical composition and the mechanical properties.

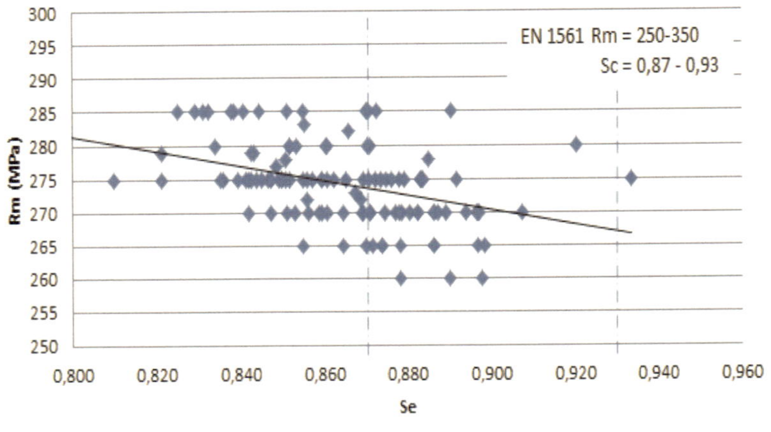

Figure 5.
Influence of the degree of saturation on the tensile strength Rm.

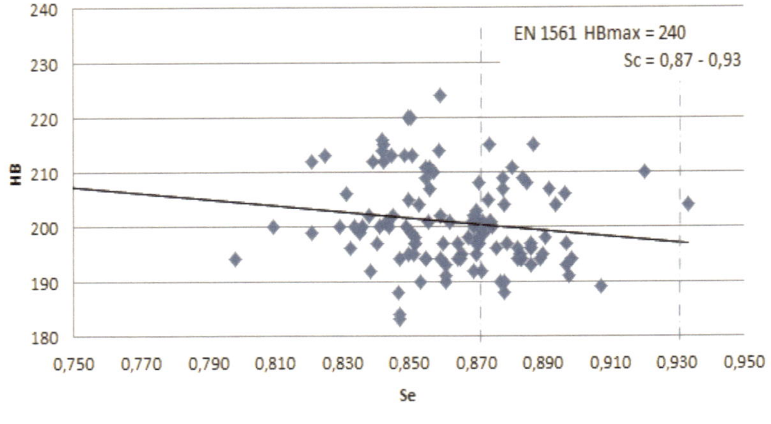

Figure 6.
Influence of the degree of saturation on the Brinell hardness HB.

Tensile strength (MPa):

$$Rm = 340.53 – 17.15\%C + 19.7\%Mn – 10.68\%Si – 34.23\%P + 10.37\%S – 57.76\%Cr \quad (12)$$

correlation coefficient 0.5849.
deviation s = 51.922.
significance level α = 0.05.

Brinell Hardness HB:

$$HB = 284.21 – 19.3\%C – 31.17\%Mn + 6.49\%Si – 19.98\%P – 11.66\%S – 31.08\%Cr \quad (13)$$

correlation coefficient 0.3852.
deviation s = 69.118.
significance level α = 0.05.

Regression dependence of the chemical composition on tensile strength Rm and Brinell Hardness HB identically show the most significant effect of content C, Mn, and Cr with literature data [16, 17].

The influence of sulfur is interesting, which increases the strength and reduces the hardness in evaluated melts.

The content of phosphorus, normally not evaluated, is significant in evaluated melts—it reduces both tensile strength Rm as well as Brinell Hardness HB.

Besides the standard evaluation of quality for cast iron by tensile strength and hardness eventually by chemical composition, the quality criteria were also evaluated (**Table 2**). It was the criteria that quantitatively represent the effect of particular production conditions on reached cast iron properties in comparison with the optimal statistically valid conditions. Really quality cast iron should show the highest degree of maturity RG and the lowest relative hardness RH.

The quality criteria for cast iron produced from 100% steel scrap rate are shown in **Table 3**. From the quality criteria (**Tables 2** and **3**), it can be stated that produced cast irons show a lower degree of maturity, which means they have lower strength than corresponds with their chemical composition. On the other side, synthetic cast iron shows a high relative hardness that in the final result decreases its quality number.

From the identified quality criteria, it can be concluded that produced gray irons show the average values of quality, while the degree of maturity RG is higher (114.5%) and the relative hardness RH is slightly lower. It is the result of the composition of the charge (without pig iron) as well as the result of low overheating of cast iron (1420–1440°C), but it is recommended to overheat it above 1480°C (for economic reasons). Despite these results of the quality criteria, it can be concluded that produced gray iron complies with EN.

The evaluated quality criteria	
The degree of maturity of cast iron RG (%)	114.5
The relative hardness RH	0.91
The quality factor m	1.363
The quality number GZ	125.8

Table 2.
The quality criteria.

The evaluated quality criteria	
The degree of maturity of cast iron RG (%)	82.57
The relative hardness RH	1.469
The quality factor m	0.771
The quality number GZ	56.21

Table 3.
The quality criteria of synthetic gray iron.

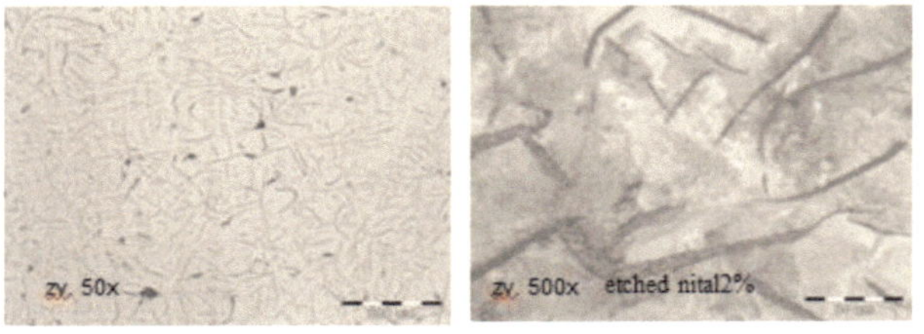

Figure 7.
Typical microstructure of gray iron of standard melts.

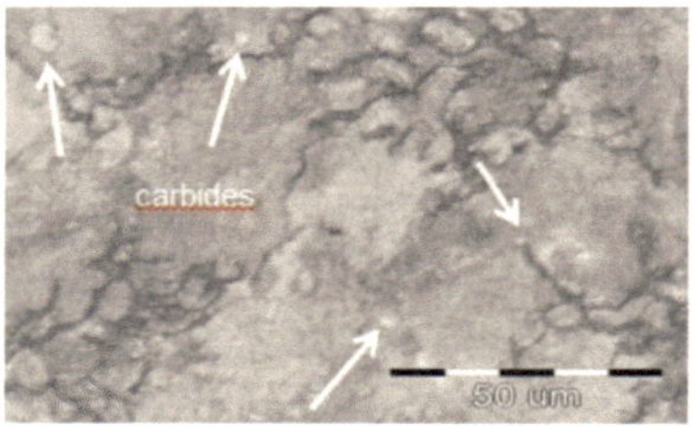

Figure 8.
Microstructure of synthetic gray iron, etched 2% Nital, 500x.

The microstructure of all melts was pearlitic with a 92–96% portion of pearlite (**Figure 7**). Cementite in the structure of melts was not observed.

In synthetic gray iron (100% steel scrap), there was full pearlitic microstructure and there were detected carbides (**Figure 8**). Attendant of these carbides was the cause of higher hardness in this gray iron.

Besides the significant influence of the chemical composition of gray iron, which is necessary to evaluate more complex, that is, including desirable and undesirable additives—chemical elements and impurities, the quality of cast iron is closely related to the metallurgy of its production. These include the following:

- quality of raw materials for a batch,
- type of furnace,
- smelting process (metallurgical processes),
- out of furnace treatment of liquid metal, and

- casting temperature.

The highest quality and the most widely used batch material for the smelting of iron is pig iron. Its quality is highly variable; it is linked up with the manufacturer, that is, the conditions of the blast furnace production. Most often, it is contaminated with trace elements and different gas content (N, H, and O), or their compounds (impurities). It is necessary to measure the content of Ti (up to 0.15%), Ca (up to 0.025%), Al (up to 0.004%), and Ni, Cr, Cu, V at its choice.

Besides the common types of foundry and steel industry pig irons, there are now also produced synthetic and semi-synthetic types that are characterized by their specific properties. It is important to check the quality of the new brand (new manufacturer) and to use one type of pig iron for a long time.

Another, particularly cost-effective batch material is steel scrap, and its quality is very diverse. Its use in the greater proportion is currently at smelting in electric furnaces (EIF and also EAF)—up to 100% share in the production of synthetic iron. The properties of such cast iron are significantly different from the properties of cast iron produced from the cupola furnace (max. 30% of steel), respectively; cast iron is produced with a higher proportion of pig iron.

It is necessary to know the quality (purity) of other batch materials, especially scrap iron, reversible material, and various impure iron-containing materials and their impact on the final quality of cast iron. Affordable material is cast iron turnings, without negative effect on the quality of the cast iron, especially the roughing of own castings.

Metallurgical processes and therefore the quality of the resulting cast iron are related to the type of furnace. The most important metallurgical parameters of smelting are temperature of the overheated metal, its standing time after melting-down, as well as the possibility of some metallurgical processes during smelting (desulfurization, dephosphorization, reduction of gases, etc.). The most extensive metallurgical treatment options are in the electric arc furnace. However, the procedure is more energy demanding.

Today, there are extensive options available (procedures and equipment) for desirable out-of-furnace treatment of liquid cast iron, which not only increases its quality, but also the cost of its production.

The general rule is to keep the lowest temperature of the casting, which is most closely connected with the wall thickness and weight of the cast and with the liquidus temperature of casting cast iron. Low casting temperature positively affects crystallization (formation of macro- and microstructure) and uniformity of characteristics of the cross section of the walls and various parts of the cast iron and, especially, it reduces the tendency to form micro- and macroshrinkages as well as heat stress during solidification.

4. Conclusions

The production of high-quality cast iron according to EN-GJL-250 (EN 1561) with a lower level of $Sc < 1$ is possible mainly in the production of thick-walled castings with a guaranteed higher required hardness. Many factors affect the quality of cast iron. Their significance is still being investigated, often with quite unexpected results (sulfur content in gray iron). For an easy assessment of the metallurgical quality evaluated by mechanical properties, statistical methods are used, especially multiple correlation and comparison with the achieved optimal parameters.

Quantitative relationships between chemical composition and tensile strength Rm and hardness HB were determined using statistical methods (multiple correlation regression equations). The following conclusions follow from the results of operational smelting in the foundry:

- C, Mn, and Cr have the most significant influence on tensile strength Rm and hardness HB.

- the quality of EN-GJL 250 (EN 1561) cast iron was mainly ensured by the reduction of Sc, which is related to the production of heavy, thick-walled castings with a guaranteed higher hardness on the walls of the casting (microalloying Cr).

- with a decrease in Sc, almost all foundry-technological properties of gray cast iron deteriorate, especially run-in and tendency to shrinkage and formation of chill-out, respectively, hard places in the corners and thin walls of the casting, which impairs their machinability. This effect was recorded in the foundry.

From the quality criteria, it can be concluded that the produced cast iron shows a lower level of RG; that is, they have lower strength than corresponds to their chemical composition. Synthetic cast iron (100% steel scrap), on the other hand, exhibits high relative hardness, which ultimately lowers its quality number (GZ 56.21).

It is a result of low overheating of cast iron (1420–1440°C); it is recommended to overheat up to 1480°C.

The desired final quality of gray cast iron is closely related to the customer's requirements, as well as to the financial costs of securing it. The difference in costs for the production of up to 100% synthetic LLG is the most current (up to €200/t).

Acknowledgements

This work was supported by the Scientific Grant Agency of The Ministry of Education of the Slovak republic No. KEGA 018TUKE-4/2022, KEGA 004TUKE-4/2023, VEGA 1/0002/22, VEGA 1/0265/21, and APVV-19-0559.

Author details

Peter Futas* and Alena Pribulova
Technical University of Kosice, Kosice, Slovakia

*Address all correspondence to: peter.futas@tuke.sk

References

[1] Pollack H. Fe-Fe3C Phase Diagram. Materials Science and Metallurgy. 4th ed. Hoboken, New Jersey, United States: Prentice-Hall; 1998

[2] Ihm M. Introduction to Gray Cast Iron: Brake Rotor Metallurgy. Livonia, Michigan, United States: TRW Automotive; 2013

[3] Collini L, Nicoletto G, Konečná R. Microstructure and mechanical properties of pearlitic gray cast iron. Materials Science and Engineering. 2008;**488**:529-539

[4] Kotus M, Andrassyova Z, Cico P, Fries J, Hrabe P. Analysis of wear resistant weld materials in laboratory conditions. Research in Agricultural Engineering. 2011;**57**:74-78

[5] Modern Casting. 2021. Available from: http://cdn.coverstand.com/55001/687958/3ba370da37f7b6c13efcbf97a5335c779891062b.4.pdf

[6] Futáš P, Pribulova A, Pokusova M, Junakova A. Liatiny—Vlastnosti, výroba, použitie, Košice. 2020

[7] Singh R. Cast iron metallurgy. Materials Performance, Materials Selection & Design. 2009;**9**:58-61

[8] Pereira A, Boehsa L, Guesserb W. The influence of sulfur on the machinability of gray cast iron. Journal of Materials Processing Technology. 2006;**179**:165-171

[9] Kumar B, Singha A. Effect on the mechanical properties of gray cast iron with variation copper and molybdenum as alloying elements. International Journal of Engineering Research & Technology. 2014;**3**:81-84

[10] Baron AA, Palatkina LV. Choice of the optimum criterion for estimating the strength of gray cast iron using the primary structure parameters. Russian Metallurgy. 2021;**5**:1555-6255

[11] Sujana W, Widi KA, Rahardjo T, Prihatmi T, N. The ability of nitrogen atomic absorption in the formation of iron nitride on flake structure and nodule in cast iron. In: Annual Conference on Science and Technology (ANCOSET 2020) IOP Publishing Journal of Physics: Conference Series 1869. Malang, Indonesia: 2021

[12] Abhijit Ramesh Patil PN. Gore: Study of factors affecting tensile strength of gray cast iron. Journal of University of Shanghai for Science and Technology. 2021;**23**(11):759-768

[13] Lian XT, Zhu JN, Dong H, Wang YM, Liu JD. Effects of micro-alloying elements on microstructure, element distribution and mechanical properties in gray irons. International Journal of Metalcasting. 2020;**14**(4):1025-1032

[14] Döpp R. Beitrag zur Entwicklung der Eisengussdiagramme. Giesserei Rundschau. 2013;**60**:328-334

[15] Mertler C, Vannatta R, LaVenia K. Advanced and Multivariate Statistical Methods. Routledge, New York: 2021. DOI: 10.4324/9781003047223

[16] Futas P, Pribulova A, Fedorko G, Molnar V. Influence of steel scrap in the charge on the properties of gray iron. ISIJ International. 2017;**57**(2):374-379

[17] Baron AA, Palatkina LV. Relation between the hardness and the strength of gray cast iron with allowance for the structural parameters. Russian Metallurgy (Metally). 2017;**11**:984-988

Chapter 6

Refractories for the Cast Iron Melting

Prasunjit Sengupta

Abstract

Refractory is a very important component in economically successful melting of cast iron. Refractory is used to line the furnace or any other vessel used for melting or holding of the molten metal. This chapter has discussed the different type of furnaces used for the melting of cast iron, the special features of those furnaces and the operational parameters of those furnaces with special emphasis on the coreless induction furnace, which is most commonly used. It has dealt with the installation practices of the refractory lining and also has discussed the mode of failure of the refractory lining and the precautions to be taken during installation and during use.

Keywords: cupola, induction furnace, quartz, silica, corrosion, buildup, superheating

1. Introduction

Cast iron is an important engineering material and has numerous applications in Civil Engineering, Architecture, Agriculture, etc. Manufacturing of Cast iron products is comprised of first melting and then casting process, normally followed in Cast iron foundries. Cast iron contains about 2–4% C and 1–2% Si and the melting takes place at a temperature of 1550C. The furnaces used for melting the Cast iron need to be lined internally with Refractory material which can withstand that temperature continuously in presence of molten metal and slag to protect the integrity of the furnace structure. Different type of furnaces can be used for melting the Cast iron and the quality of the Refractory material used depends upon the furnace type. Although the Refractory cost per ton of processed cast iron is not high and significant but the refractory plays a very vital role towards the economic viability of the project.

The Refractory lining influences the furnace operational efficiency and the productivity. Higher is the refractory life, lower is the stoppage time and higher is furnace availability and productivity. Refractory quality influences quality of the castings also, because of the inclusion, coming from refractory. The incorporation of inclusion in the casting impairs its mechanical property and appearance.

We shall discuss here about the different types of furnaces used for the cast iron melting, different types of the Refractories used for different kind of furnaces, installation of refractories and the cause of failure of Refractories.

IntechOpen

2. Furnaces used for melting cast iron

Three types of furnaces are mostly used for the Cast iron melting in the foundries. These are:-

1. Cupola Furnace

2. Induction Furnace

3. Arc furnace

The refractories are designed to withstand the operational conditions and the environment inside the furnace and therefore it is essential to know the furnace and operational details connected to refractory selection.

2.1 Cupola furnace

The typical Cupola furnace cross sections are shown in **Figure 1** [1].

The cupola is a shaft type cylindrical structure as shown in **Figure 1**. The principle of operation is similar to Blast furnace. The pig iron and scraps, lime stones and coke are charged from the top and air is blown from the bottom when the oxidation of Carbon generates heat which melts the iron and it is collected at the bottom of the furnace from where it is tapped out. The hot air moves upward through the bed exchanging its heat with the downward moving cold burden and preheating it. Two major reactions take place inside the cupola. The first one is the oxidation of carbon in the coke $C + O_2 = CO_2$, which is highly exothermic reaction and increases the temperature inside and melts the iron. The molten iron picks up the carbon according to the reaction $3Fe+2CO = Fe_3C + CO_2$ which is an endothermic reaction. Temperature

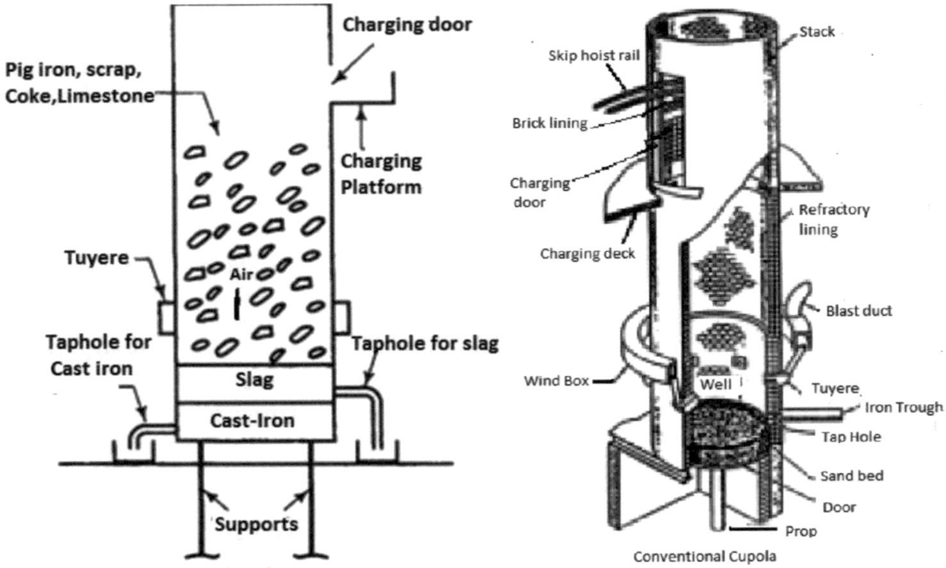

Figure 1.
Cupola furnace.

DOI: http://dx.doi.org/10.5772/intechopen.105973

inside the Cupola can reach up to 1600C and temperature control is not easy and precise in the Cupola furnace. Cupolas are used mainly where the large volume of Cast iron is melted.

2.2 Arc furnace

The typical arrangement of Arc furnace is shown in **Figure 2** [1]. This furnace is a cylindrical Refractory lined steel shell fitted with three Carbon electrodes inserted through the Refractory lined roof of the furnace. The scrap and the pig iron and the fluxes are charged through the door inside the furnace and the power is made on. The arc between the charge and the electrodes produces arcs which generates high temperature and melts the charge.

2.3 Induction furnace

This is the most widely used furnace in the foundry for melting the cast iron because of its easy and precise control of temperature and melt chemistry. The principle of induction furnace is same as a transformer. The electric current is passed through a water cooled coil and when electrically conductive solid metal kept inside the coil, it gets heated up because of induction. Two different types of induction furnaces are there.

2.3.1 Coreless induction furnace

The coreless induction furnace basically consists of the cylindrical refractory crucible surrounded by the induction coil supported by the transformer yokes (**Figure 3**). The energy is transmitted in such induction furnaces by passing an electric current through the coil which creates a magnetic field. Voltages are induced in the feed material due to this magnetic field, whereby eddy currents are created due to the conductivity of the metal. The induced current heats the charged material in accordance with Joule's law and it melts after a certain heating time.

Any current I, AC or DC, passing through an electrically conducting material causes a voltage drop V resulting in energy conversion to heat. Heat generated in the process is defined by $V.I = R.I^2$, where R is the electrical resistance of the current path.

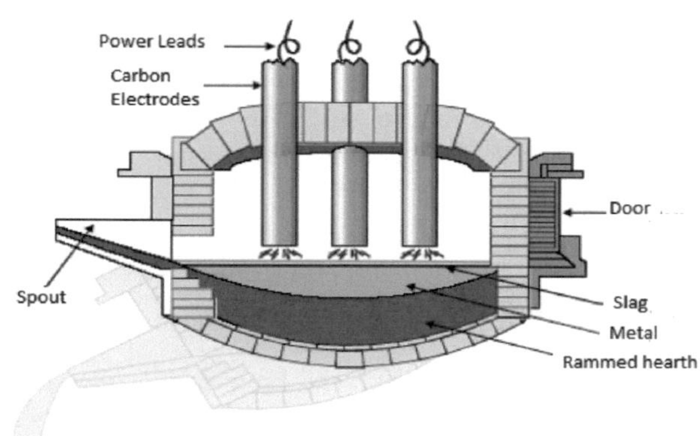

Figure 2.
Arc furnace.

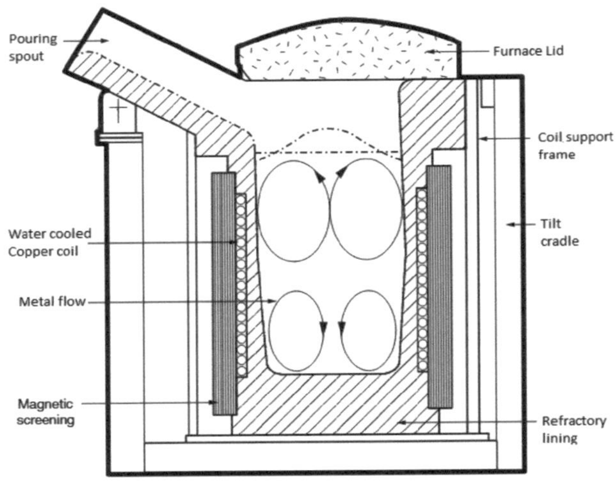

Figure 3.
Vertical section of coreless induction furnace.

The resistance of the current path is inversely proportional to the cross-section area in which the current is flowing.

Coreless induction furnace normally has capacity up to 50 Ton. It can run by the main frequency of 50 Hz or by frequencies 200-1000 Hz called medium frequency furnace. Medium frequency furnaces are mainly used in industries because they have got certain advantages over the main frequency furnaces e.g.,

1. It requires less energy to melt the same quantity of metal.

2. It does not require to maintain any metal heel inside the furnace.

3. The furnace can be emptied and therefore no chance of mixing in case of different metals are produced in the same furnace.

4. No necessity to use the starter block to start cold heat.

2.3.2 Channel induction furnace

Channel induction furnaces are mostly used for holding the molten metal or for superheating the molten metal. It is used along with the Cupola to hold molten metal. The arrangement of the furnace is shown in **Figure 4**.

Power consumption – Theoretically, melting one ton of Cast iron at 1500C should consume 396 Kwh of energy, but in actual practice it takes about 500 Kwh of energy because many types of energy losses takes place which is shown in **Figure 5** [2].

2.4 Comparison of the furnaces

All the different types of furnaces discussed have some special features, advantages and disadvantages. Selection of the right kind of furnace depends upon the grades of metal to be produced, required melting capacity, availability and cost of raw materials and consumables, cost of electricity and coke, capital investment, operational cost, environmental restrictions, available space etc. Amongst all these different

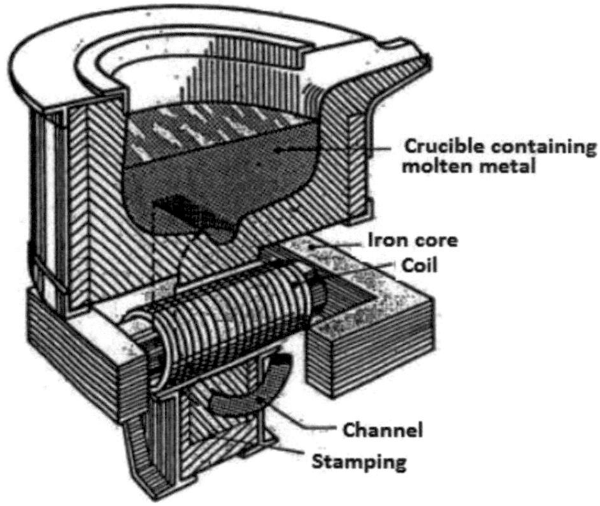

Figure 4.
Channel induction furnace.

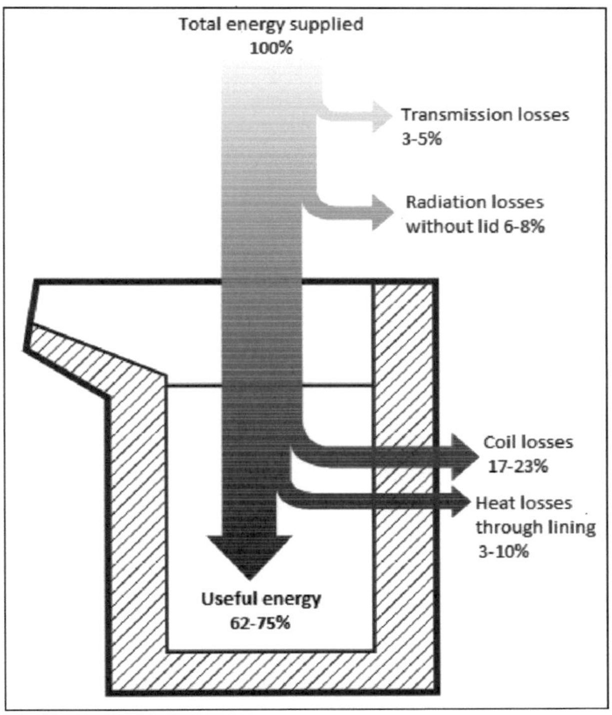

Figure 5.
Energy balance in the coreless induction furnace.

type of furnace used for Cast Iron melting, most used furnaces are Cupola and Coreless Induction furnace. The comparison of the special features of Cupola and Coreless Induction furnace is shown in **Table 1** [3, 4].

Features	Cupola	Induction furnace
Energy consumption/ton of metal (KWh)	825–890	490–520
Efficiency%	44–47	75–79
Energy Consumption/ton of metal (Auxiliary Equipment) (KWh)	20–70	6–10
Throughput (Ton/h)	5–100	0.1–50
Temperature variation (K)	20–50	5
Chemistry control		
Delta C%	0.5–0.7	0.1
Delta Si%	0.5–1.2	0.1
Slag quantity (Kg/ton)	40–100	10–30
Dust quantity (Kg/ton)	5–15	0.1–0.3
Metal losses (%)	0.5–1.5	0.1–0.3
Mode of operation	Continuous	Intermittent
Charge quality	Can handle highly oxidized and low quality scrap	Require high quality scrap
Maintenance cost	Higher	Lower
Capital investment	25% higher	25% lower
Refractory cost	Lower (1.8USD/t)	Higher(3.1USD/t)
Flexibility	Limited	Very high
Refractory lining	Complicated	Easy
Temperature homogeneity	Good	Very good
Chemistry adjustment	Good	Very good

Table 1.
Comparison of features of cupola and induction furnace.

3. Refractory

Materials are inert inorganic solid materials which can withstand high temperature in contact with solid, liquid and gases to retain its integrity and mechanical strength. These are basically Oxides, Nitrides, Carbides and Borides of Aluminum, Silicon, Alkaline earth metals and transition metals. Selection criteria of refractory for a high temperature process depend mainly upon the following.

1. Operational temperature

2. Chemical constituents, it will be in contact with and their state of occurrence at that temperature (solid, liquid or gas)

3. Batch process or continuous process

4. Mechanical abuse by the solid charge, degree of turbulence of liquid or gas velocity

5. Thickness of the refractory lining and shell temperature permitted as per design

6. Furnace size and geometry.

7. Economic considerations

Refractory lining has primarily two main functions

1. To withstand the hostile environment inside the furnace to protect of the furnace steel structure.

2. To reduce the outward flow of energy from furnace inside to conserve energy.

3.1 Cupola

It is to be borne in mind that there is lot of difference between a modern Cupola and the Cupola of earlier days. In earlier days Cupola used to be run for a day and next day it was used to be put down for Refractory maintenance and slag cleaning. Very low quality patching material comprised of sand, burnt brick bats and clay mixture were used for patching purpose. But to-day the modern Cupola is designed for continuous running and therefore high quality Refractories are used to line inside. Cupola is a vertical shaft furnace in which different operational condition exists across its height (zone) and different quality of refractories are used at different zone of the furnace (**Figure 6a**) [5]. The thermal profile across different zone is shown in **Figure 6b** [6].

Charging is done from the top and the uppermost zone experiences heavy mechanical abuse and abrasion by the falling charge material like scrap, limestone and coke and descending burden.

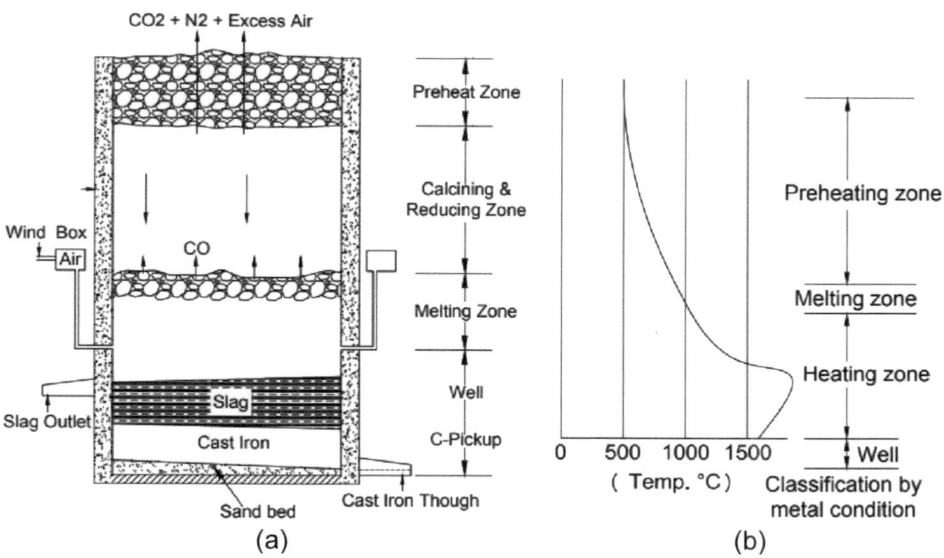

Figure 6.
a. Different zones of the cupola furnace. b. the temperature profile inside cupola.

Area of use	Al_2O_3%	SiO_2%	SiC+C%	CaO%	Bulk density (gm/cc)	Maximum Service temp (°C)
Upper stack/ Cal. zone	60–70	25–35	—	2–4	2.2–2.5	1600–1700
Melting zone	10–40	5–6	45–80	0.5–1.5	2.65–2.70	1700
Well /Hearth	45–75	5–6	20–45	0.5–1.0	2.75–3.00	1700
Taphole area	65–70	4–5	20–22	1–3	2.90–3.00	1700
Iron Runner	60–75	5–6	18–20	1–2	2.9–3.00	1700

Table 2.
Typical properties of Castables suitable for different zone of cupola.

Preheating and calcining zone are heated by the upward moving gas and the temperature there is low, below 1000C (**Figure 6b**). High density and low porosity 60–70% Al2O3 refractory bricks or dense low cement Castable can be used here for lining.

In the melting and well zone high quality Al2O3-SiC-C refractory either in brick form or as low cement self- flow Castable is used. Corundum based Castable or bricks can also be used here.

The tuyere is lined with high strength Al2O3-SiC-C or Corundum based low cement Castable. The refractories are degraded mostly by the chemical corrosion by the fluid slag generated in the melting process. The primary condition for the corrosion is the wetting of the refractory surface by the molten slag. The wettability of a solid surface by a liquid happens when the contact angle is low. The addition of Carbon to the refractory increases the contact angle of slag to refractory surface to make it non-wetting and hence more corrosion resistant [7]. The thickness of the lining depends upon the diameter of the Cupola and the intended shell temperature. To maintain low shell temperature and lower refractory thickness the use of insulation refractory is inevitable, otherwise the water cooling of the shell is required.

The use of water cooling leads to energy loss and increases the fuel cost.

Chemical and physical properties of refractory castables used for different zone of the cupola furnace are shown in **Table 2**.

3.2 Coreless induction furnace

The Refractory issue is much more critical in case of Coreless induction furnace. In Coreless induction furnace the refractory lining separates the molten metal from the electrical copper coil behind (**Figure 3**) through which water is circulated to keep it cool.

The refractory lining thickness is to be optimized taking into consideration of the following:-

1. The higher is the thickness better it is to prevent heat loss through the lining.

2. Higher is the lining thickness lesser is the electrical efficiency of heating (**Figure 7**) [3].

3. Safety is a very important issue to be looked into during the operation of induction furnace. If a crack forms in the lining during the operation of the furnace, the molten metal may penetrate inside and strikes the coil which may lead to severe explosion.

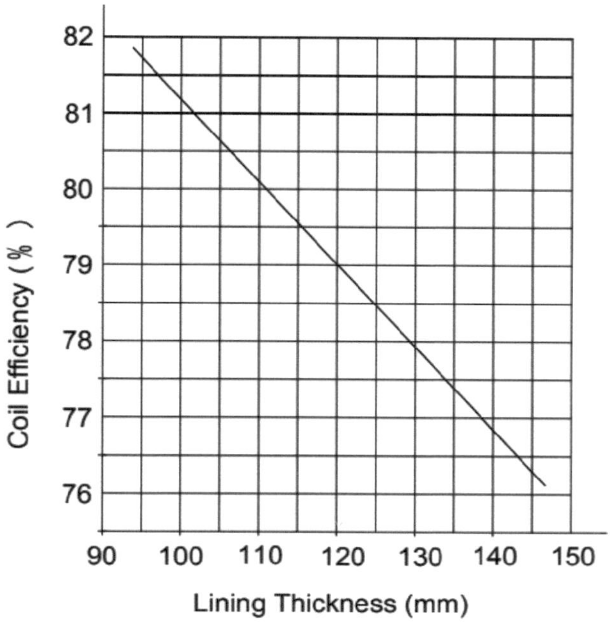

Figure 7.
Effect of refractory lining thickness on electrical efficiency.

Thinner the lining more is the chance of metal penetration. On the other hand, higher is the refractory lining thickness lesser will be the heat loss through the lining but more will be the loss in electrical energy input given, to heat the charge inside the furnace.

Decreasing refractory thickness improves the coil efficiency but at the same time admits higher thermal losses through the thinner crucible wall. However, since coil losses exceed the thermal losses across the crucible wall nearly by the factor of 10, coil losses play the dominant role here. Taking above all points into consideration high safety margin is eliminated by use of advanced crucible monitoring equipment and the lining thickness is optimized.

The refractory lining thickness commonly maintained is from 75 to 125 mm based on the furnace capacity. The concept of refractory lining design for the induction furnace is quite different from conventional refractory lining.

The main attention is given to arrest crack formation and crack propagation, across the lining thickness, which may pose the danger of penetration of molten metal to strike the coil. During the operation of the furnace the refractory lining always experiences temperature fluctuation and related thermal shock, because the furnace does not run continuously at same temperature. The thermal shock generates crack in the lining, therefore, lining by brick or ramming masses of conventional type are not suitable because in sintered refractory brick, in castable or in a chemically bonded ramming mass, if a crack forms, it propagates very fast.

One of the measures taken to handle this problem is to use the dry powdery mass for lining. The material is designed such that during the use only one third of the lining thickness at the working face will get sintered hard which will withstand the chemical and mechanical abuse of molten metal and charging scrap. The next one

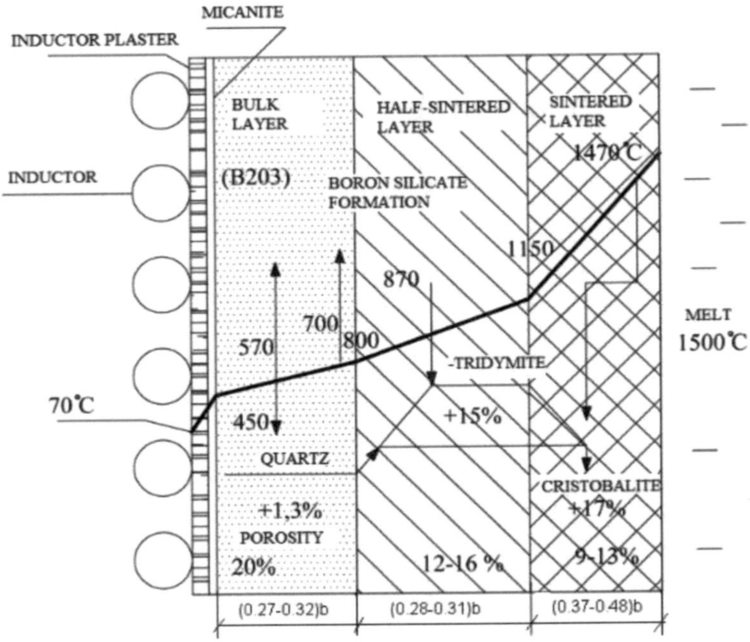

Figure 8.
The thermal profile across the refractory lining thickness in coreless induction furnace (b is total thickness).

third of the thickness will be in semi-sintered stage and the rest one third of the thickness at the back, in touch with inductor coil, will be in loose form (**Figure 8**) [8]. Under such condition if a crack forms at the working face, it cannot propagate up to the coil and will be arrested in between. The diagram also shows the temperature profile of the lining in case of Quartzite lining (Silica Ramming Mass) which is mostly used in case of Cast iron melting.

The requirements of a ramming mass suitable for the lining of a coreless induction furnace are

1. It should have a softening point minimum 150C above the operating temperature

2. It should sinter in working layer to have a strength to withstand the pressure of molten metal.

3. Should have a poor sinter-ability not to allow sinter at back

4. Should be chemically inert at the operating temperature in contact with the molten metal and slag.

5. Should have permanent expansion at a temperature 1000-1200C

6. Should have a low thermal conductivity

7. Should be economic

3.2.1 Silica ramming mass

Meets all the criteria mentioned before, to make it a most commonly used lining material for the Coreless induction furnace for melting Cast iron. It is made out of high purity Quartzite or Quartz containing minimum 98.5% SiO2. 0.2–1.5% Boron containing compounds like Boric acid or Boron oxide is used as the sintering aid and mixed with the ramming mass during supply. The percentage of these additives depends upon the operational temperature of the furnace. The grain size distribution in the ramming mass is of utmost importance. The grain size distribution determines its packing density on compaction. Higher is the packing density better will be the performance.

Quartz the main mineral phase in this ramming mass can exist in different crystalline phases and undergoes polymorphic changes at different temperature as shown in **Table 3.**

Due to this polymorphic transformation and associated volume change from Cristobalite to Tridymite the Refractory lining at the middle layer remains tight and does not allow any crack to proceed further and stops any liquid metal penetration. Chemical and physical properties of Silica Ramming mass used for the lining of Induction furnaces are shown in **Table 4.**

3.2.1.1 Installation

Installation is a very important part towards the efficient running of the furnace. The campaign life of the lining and the electrical efficiency depends upon the quality of the installation. Best quality refractory material will not produce the desired performance unless the installation is sound. **Figure 9** [3] shows how the electrical efficiency is related to the packing density. The aim of good installation is to get maximum and uniform packing density. The process of packing the Refractory mass between the coil and the central former can be done both manually as well as through mechanization. For smaller furnaces below 5 ton capacity manual ramming may be done but for larger furnaces mechanical ramming method must be adopted to ensure best and uniform packing.

Change in crystalline phase	Transformation Temperature (°C)	Associated volume change%
β to α- Quartz	573	+0.8–1.3
β Quartz to β- Tridymite	870	+14.4
β Quartz to β-Cristobalite	1250	+17.4

Table 3.
Polymorphic changes in silica relevant to function of silica ramming Mass.

SiO2%	Al2O3%	Fe2O3%	Others%	Bulk density (gm/cc)	Max service temp C
99.2	0.5	0.1	0.2	2.1	1650
98.8	0.7	0.07	0.40	2.1	1700

Table 4.
Properties of typical silica ramming Mass.

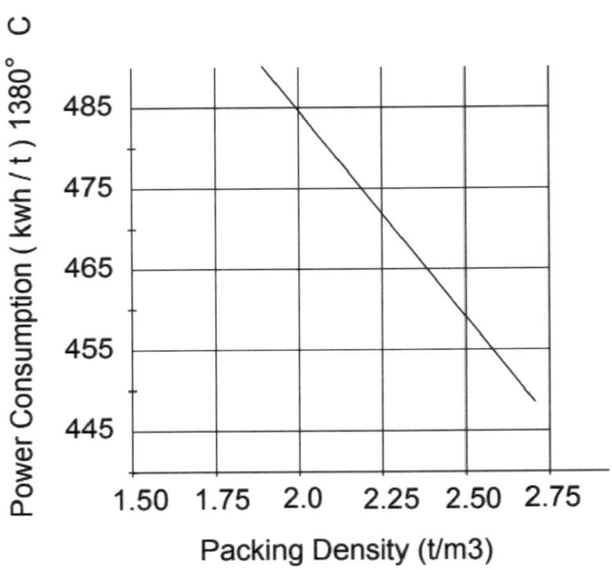

Figure 9.
Relation between the packing density and the energy consumption (MFT-GE/10000/8000KW/250 Hz).

It is necessary to check, before installation that the ramming mass does not contain any moisture and is perfectly dry. It is very safe to heat the ramming mass before installation to ensure the removal of any moisture. In actual installation process the ramming mass poured on the bottom is first rammed to compact. The ramming material should be poured in such a way that every time sufficient material is poured to get a compacted height of 50 mm and gradually the desired height is build up. The material must not be poured from a height which may segregate the coarse and fine particles of the mass. In case of pouring from a height, a long funnel must be used to ensure no segregation. After the bottom is ready its level is checked and then a mild steel former of the shape, shown in **Figure 10a** is placed on the packed bottom and the annular space between the coil and the former outer wall is packed with the Ramming mass. **Figure 10b** shows some of the tools being used for the compaction of the refractory ramming mass.

Ramming is done dry and therefore is difficult to compact and the tools are also of different design than those used for wet ramming.

Before ramming the inside wall of the coil is plastered with mixture of fine Alumina powder and high Alumina cement mixed with water. The presence of coarse grains may damage the Copper coil. The coating thickness will be 3–5 mm. This provides an extra layer of protection over the coil and also forms a separation layer between the coil and the ramming mass which makes the removal of the used up lining easier after the campaign life of the ramming mass is over.

There is a practice to put some insulation layer like Asbestos sheet over the coil to thermally insulate the coil, but this is not recommended as a correct practice. It consumes both money and time and reduces the lining life by helping sintering front to move towards the coil. This is against the philosophy of the lining design of the coreless induction furnace and moreover Asbestos creates health hazard.

After completion of lining, the scrap is charged inside and the power is put on. The steel former used for the lining is allowed to melt in the process of sintering the lining.

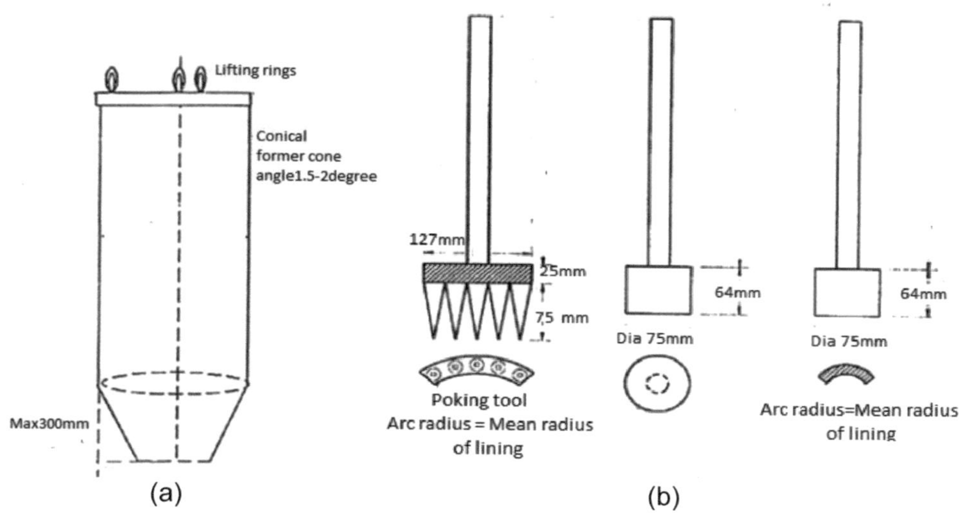

Figure 10.
a. Steel former. b. Ramming tools.

There are methods in which the steel former can be taken out after the ramming is over and before the power is made on. This process is much more economical because the same former can be used number of times and it saves the cost of the former. For removing the former before melting, special binder is added in the ramming mass and the former is heated up to 400C by gas burner inside when the ramming mass, in contact with it, forms a hard layer and enables the pull out of the former which has a taper design to facilitate the removal.

The first heat is very important because it is to be done with care to sinter and stabilize the lining and next heat onwards it can be run in normal routine way. It is also recommended to use clean and good quality scrap in first few heats. The former used for the lining is allowed to melt in service and it actually holds the dry material till it sinters and acquires strength to stand on its own.

After a new lining is constructed, its first heat up procedure is very important. A special heat up schedule is followed which helps in stabilization of the Refractory lining. This is called sintering cycle and should be followed as per the instruction of the supplier of the lining material because it depends upon the grain size distribution, raw material character, quality and quantity of the sintering aid used and also upon the furnace capacity. The most common sintering aid is Boric acid or Boron Oxide used for Silica Ramming mass. **Table 5** shows the typical heat up schedule used for a certain ramming product.

After rising, the temperature is hold at a certain temperature, called sintering temperature, which depends upon the quantity of additive used. **Table 6** shows the relation between the kind of sintering aid used, its percentage and the sintering temperature.

3.2.1.2 Mechanism of refractory degradation

Refractory lining life depends not upon quality of the refractory alone but more upon the other parameters e.g., furnace size, quality of installation and the

Furnace size (Ton)	Boron oxide (B2O3)	Boric acid (H3BO3)
1–3	180 C/h	120 C/h
4–15	150 C/h	100 C/h
+15	100 C/h	60 C/h

Table 5.
Heat up schedule of the coreless induction furnace.

% Additive	B2O3	H3BO3
1.0	1530C	—
0.8%	1580C	—
0.6%	1630C	—
0.4%	1680C	—
2%	—	1530C
1.6%	—	1580C
1.2	—	1630C
0.8%	—	1680C

Table 6.
Typical relation between additive% and sintering temperature.

deviations from SOP etc. Same Refractory performs differently in different cast iron melting units.

Following are the major causes of refractory degradation process takes place in induction furnace.

1. Chemical corrosion and erosion
2. Crack formation
3. Erosion
4. Superheating
5. Build up

3.2.1.2.1 Chemical corrosion and erosion

It causes the gradual loss of the lining thickness due to the chemical reaction of the refractory material with the charge material or alloying elements in the metal. The presence of carbon and other oxides like Mn, Mg, and Al present in the melt reacts with SiO2 and reduces it following the reaction as below.

$$2X + SiO2 = 2XO + Si \tag{1}$$

The content of the said impurities must be low to avoid the chemical corrosion. Carbon in Cast iron also attacks SiO2 at a temperature above 1450C to 1480C when the boiling process sets in. The FeO present in the slag reacts with SiO2 to form low melting compound Fayalite (2FeO.SiO2) having melting point 1180C and corrodes SiO2 Refractory lining. Slag also contains Manganese silicate (MnO.SiO2) with 1250°C as its melting temperature. Both of these compounds occur in the slag in proportion 10 to 30% and 2 to 10% respectively. Input of rusty scrap makes the situation worse.

The content of the said impurities must be low to avoid the chemical corrosion. Carbon in Cast iron also attacks SiO2 at a temperature above 1450C to 1480C when the boiling process sets in.

During the charging of the scrap in molten metal or during the CO boiling process, molten metal is splashed on the refractory lining and later gets oxidized and forms FeO and then to Fayalite.

When the metal is hold in the furnace for longer time metal gets oxidized to FeO and causes the erosion in the lower part of the crucible. During holding, the temperature of the molten metal should be kept as low as possible to retard the oxidation process and generation of FeO. In melting of nodular iron, SiO2-Al2O3-MgO eutectic is formed at 1365C [9] and the lining gets eroded because if it's higher tapping temperature.

If the scrap, used as feed to furnace, contains Zn, then it vaporizes beyond 900 C and permeates through the lining material and condenses on the inductor coil. This deposited Zn layer may cause arc formation in the inductor coil and damages it [8]. To avoid this problem the initial three charges must be free from any Zn and the lining is allowed to sinter and get dense to retard the permeation of Zn vapor. The same is applicable during the charges after cold heat because the time is to be allowed to heal up the cooling crack through which the vapor permeates easily. A high temperature gradient from hot to cold face of the lining is also recommended to allow the condensation zone away from the coil. The coil coating material with high thermal conductivity is recommended.

The CO gas also permeates through the lining and gets converted to C and CO2 as per the Boudouard reaction

$$2\mathrm{CO} = \mathrm{CO2} + C \tag{2}$$

This Carbon gets deposited on the coil. Beside this, Sulfur vapor generated from the MgS in recycled nodular iron also penetrates through the lining and reacts with the Oxygen and the moisture in the lining to form Sulfuric acid as per the reaction

$$S_2 + 3\mathrm{O}_2 + 2\mathrm{H}_2O = 2\mathrm{H}_2\mathrm{SO}_4 \tag{3}$$

Sulfuric acid attacks the Copper coil to form Copper Sulfate and damages the coil.

3.2.1.2.2 Crack formation

in the lining is inevitable because of thermo-mechanical stresses developed in heating and cooling of the lining and due to volumetric changes during the polymorphic transitions of Quartz. But formation of deep crack is dangerous which may allow the passage of molten metal through it to strike the coil. Formation of small cracks is not a concern rather formation of smaller cracks absorbs the stress and does not allow the formation of bigger cracks. Cracks can be of different type and the reason of their formation is shown in **Figure 11** [10].

The lamination is formed due to separation of two layers during compaction of the lining and care must be taken to avoid such layer formation. The vertical cracks are

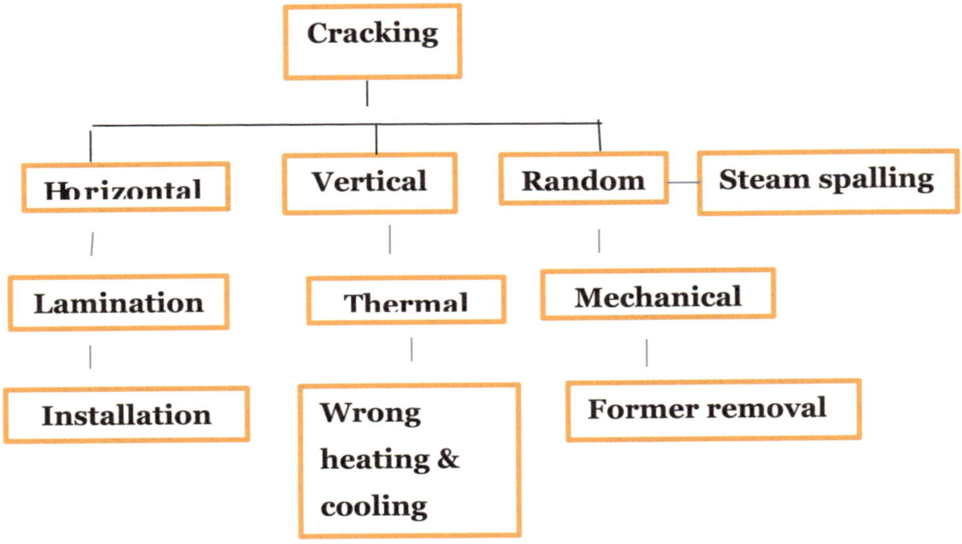

Figure 11.
Different types of lining cracks and their causes.

formed during over sintering also and care must be taken to reduce the amount of sintering agent in such case. Crack also formed if the density is not uniform throughout the lining. Segregation of the material can also cause such non uniformity. To avoid segregation the refractory material must not be poured from much height or to pour the material through a long funnel during lining.

3.2.1.2.3 Erosion

Erosion is more of a physical process and aggravates chemical corrosion by exposing the fresh surface available for chemical reaction and corrosion. Erosion is connected to the extent of turbulence of the bath of molten metal. Higher the turbulence more is the erosion.

The characteristic of induction melting is that the bath is in constant movement, which is called inductive stirring. The amount of stirring is determined by the size of the furnace, the power put into the metal, the frequency of the electromagnetic field and the type/amount of metal in the furnace.

When a furnace is operated at a frequency lower than ideal, the result may be a violent stirring action that may produce inclusions of slag and refractory particles. Metal loss may be excessive due to excess surface area of the melt and oxidation of volatiles.

In many cases the refractory lining life is reduced because of using too low of a frequency to produce strong stirring. On the other hand, if too high a frequency is selected for the size of the furnace, there may be a complete lack of stirring, uneven heating throughout the charge, excessive side-wall temperatures and difficulty in attaining homogeneous melts.

The degree of agitation in molten metal can be indicated by Stirring Index, which is defined as [11].

$$\mathrm{SI} = \frac{6000\sqrt{\mathrm{KW}.\frac{\mathrm{D}}{\mathrm{SG.f.}\rho}}}{A} \quad (4)$$

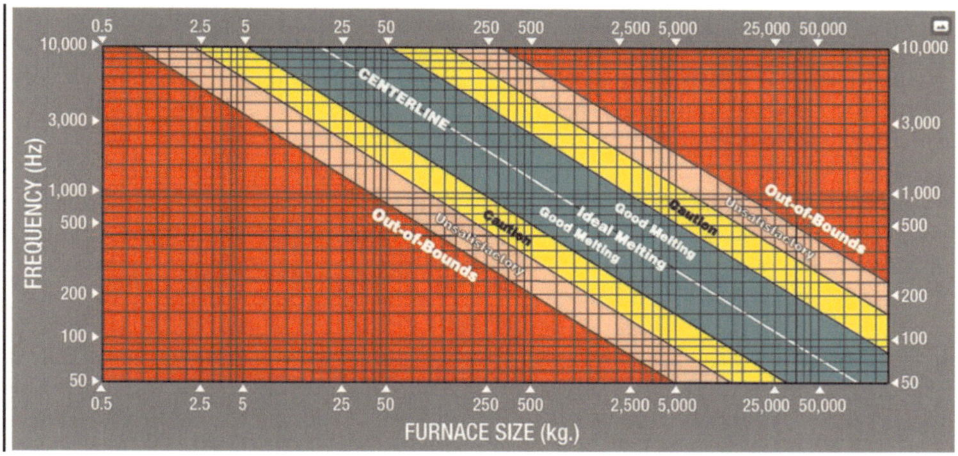

Figure 12.
Relation between the furnace size and the ideal frequency of operation.

Where, SI- Stirring Index, KW = Power of Furnace in KWh, D = Melt diameter, SG = Sp. Gravity of metal, ρ = Metal resistivity, f = Frequency, A = Cross sectional area of the melt ($\pi D^2/4$). Relation between furnace size and frequency is shown in **Figure 12** [12].

Due to erosion when the lining gets thinner, the furnace draws more power and the melting rate becomes faster and this is an indicator of lining erosion. Data in **Table 7** [2] illustrates this effect in a 3 ton capacity furnace of 700 KW rating.

3.2.1.2.4 Buildup

When the slag makes contact with the refractory lining of a furnace wall (or other areas of the holding vessel) that is colder than the melting point of the slag, the slag is cooled below its freezing point and adheres to the refractory furnace wall or inductor channel. The source of these build up material are the oxides from the oxidation of the metal or contaminants charged into the furnace e.g. molding sands. Buildup normally occurs in the areas where the flow or the turbulence is minimum. Some of the major mineral forms found in the buildup, are shown in **Table 8** [13].

The buildup gradually reduces the working volume of the furnace and forced to take shutdown for the new lining. The remedy is to use the better quality of scrap with lesser contaminants. Sometimes the use of flux reduces the build up by reacting with it to reduce the melting point so it goes into the slag.

Campaign	Power input (KW)	Energy consumption (KWh/t)
New Lining	615	656
Lining after 1 week	655	622
Lining after 3 week	750	598

Table 7.
Effect of refractory lining age on energy consumption.

3.2.1.2.5 Superheating

The generation of localized heat, which leads to high temperature at some spots, is very detrimental for the refractory lining and may cause lining failure. The major reasons for the localized superheating of the lining are shown in **Figure 13** [10].

During the scrap charging in the furnace, it may so happen that some scraps remain at hanging position at the top while the liquid metal is formed below and an air gap forms in between the liquid metal at the bottom and the charge at the top in hanging position and is called Bridging. This is a very dangerous situation for the refractory because the liquid metal below will be superheated and will have high stirring effect due to high power density and lesser quantity of molten metal. The metal temperature can shoot up above the melting point of refractory and erosion will be high because of strong agitation of molten metal. Under such condition the refractory lining can give way and molten metal can penetrate through the lining to strike the water cooled coil causing severe explosion (**Table 8**).

Once the bridging of scrap happens the power must be switched off immediately. The scrap sizes are very important to control the bridging and the charges must be of different sizes.

The trapped metal pieces inside the refractory lining can also cause the local superheating of the refractory lining. The penetrated metal fin inside the lining can also cause the superheating of the lining.

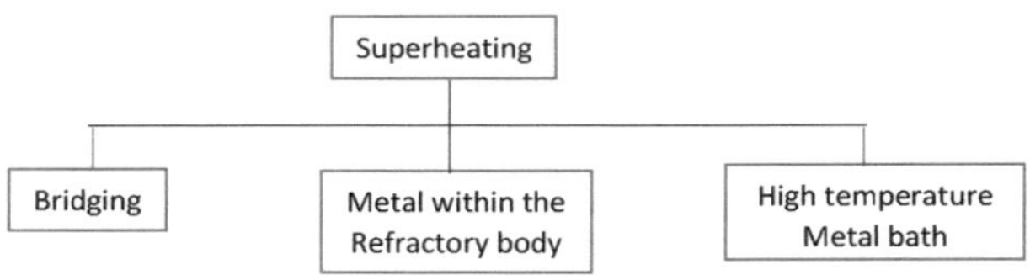

Figure 13.
Factors that cause superheating of molten cast iron.

Formula	Mineral	Melting point (°C)
FeO	Wustite	1379
Fe2O3	Hematite	1625
2MnO.SiO2	Galaxite	1850
2FeO.SiO2	Fayalite	1216
MgO.SiO2	Forsterite	1888
2(Fe, Mg)OSiO2	Olivine	1798
CaS	Oldhamite	2522.5
CaO.MgO.SiO2	Diopside	1389.7
CaO.Al2O3.SiO2	Anorthite	1555.6
MgO.Al2O3.5SiO2	Cordierite	1576

Table 8.
Some minerals, found in the buildup material in cast iron melting.

3.3 Transport of molten metal

For transportation of the molten cast iron in foundries Ladles can be used and these ladles can also be lined with Silica Ramming mass in similar way as it is being done for induction furnace. The advantage of Silica Ramming mass is its low cost and lower drop in metal temperature because of its low thermal conductivity. Smaller foundry ladles also use sol-gel castable lining which is amenable for fast drying and heat up.

In case of cupola also, the liquid metal can be transported through ladle lined by Silica Ramming mass which is most economical. In case of bigger ladle Alumina—Silicon Carbide bricks can be used which gives better campaign life but it is having much higher thermal conductivity and need insulation at the back to prevent the heat loss.

4. Concluding remarks

The driving force acting on industries, in general, today are related to economy, environment and safety and health issues and that brings the changes in current practices. The cast iron industry is not an exception to that.

For example, so far Boron compounds are being used as the sintering aid to Silica Ramming mass, but Boron compounds are found to have detrimental effects on human health. Trials are on the way to develop Boron free Silica Ramming masses and initial trial results are very encouraging.

New methods for melting, like Electron beam melting, Microwave melting, Solar furnaces [13] are under trial and Refractory requirements will be changed along with the changed furnace type in future.

Author details

Prasunjit Sengupta
SKG Refractories Ltd, Nagpur, India

*Address all correspondence to: pseng4311@gmail.com

References

[1] Naranjo RD, Kwon J-Y, Majumdar R, Choate WT. Advanced Melting Technologies-Energy Saving Concepts and Opportunities for the Metal Casting Industries. Columbia, USA: US Department of Energy; 2005

[2] Medias J. Efficient Melting of Coreless Induction Furnace – Good Practice, by Energy Efficiency Enquiry Bureau. Harwell, Didcot, Oxfordshire: ETSU. Argentina, UK. 2000

[3] Schmitz W, Trauzeddel D. The Melting, Holding and Pouring Process-Energy and Process Related Aspects. Germany: Otto Junker GmbH; 2010

[4] Medias J. Alternatives for Hot Metal Production: Cupola, Induction and Arc Furnace. Buenos Aires, Argentina: Jorge Medias; 2010

[5] Eliott R. Cast Iron Technology. Somerset, England: Butterworth & Co. (Publishers) Ltd. 1988. p. 47

[6] Takahashi T. Refractories Handbook. Japan: The technical association of Refractories; 1998

[7] Khanna R, Ekram-ul-Haq M, Sahajwalla V. Chapter-10: Influence of wettability and reactivity on the degradation-interaction of molten iron and slag with steelmaking refractories at 1550C-R. In: Wetting and Wettability. London, UK: Intech Open; 2015. DOI: 10.5772/61271

[8] Kukartsev VA, Kukartsev VV, Tynchenko VS. Causes of quartzite lining destruction during operation of the IChT furnace and the ways to prevent them. In: IOP Conf. Series: Earth and Environmental Science. UK: IOP Publisher; Vol. 459. 2020. p. 062090

[9] Dotsch E. Refractory demands on inductive melting of cast iron. Baden-Baden, Germany: Refractories Worldforum; 2011. p. 3

[10] A trouble shooting guide to Silica Dry Ram Refractories-M/S Allied Refractories, Columbia, USA, 60th Indian Foundry Congress Souvenir

[11] Prabhu S. Metal Stirring in Coreless Induction Furnace. Foundry Management and Technology; OH, USA. 2018

[12] Selecting the right unit for the efficient induction melting – Michael Fanz Huster, USA: Foundry Management and Technology; 2021

[13] Mechanism and Control of Buildup Phenomenon in Channel Induction and Pressure Pouring Furnace-Part 1, David C Williams and R.L.(Rod) Naro. Ohio, USA: Ductile Iron; 2007;**1**:44-55. Available from: https://www.asi-alloys.com/pdf/Buildup%20%20Phenomenon%20in%20Channel%20furnaces%20DIS%202007.pdf